PRAISE FOR *The Mercury 13*

"[Ackmann] has done a commendable job of documenting this story, one that deserves to be heard out of respect for the women who wanted nothing more than an equal chance, and in recognition of their courage that paved the way for the next generation of aspiring women astronauts." —Cleveland *Plain Dealer*

"In this brisk, bracing book ... Martha Ackmann takes us back to an unregretted period not so long ago when a female aviator was expected to wear a dress and high heels under her flight suit. . . . [A] stirring story." —*Newsday*

"[A] sharply pointed narrative [about how thirteen] highly skilled fliers were grounded, leaving it to the Russians to put a woman in space fully twenty years before the American government saw fit to do so. A shameful episode exposed with thoroughness and a graceful pen." —*Kirkus Reviews*

"This well-researched, elegantly written book not only pays deserved tribute to those thirteen pilots but also underscores the dismal history of sexism at the highest levels of American government." —*Chicago Sun-Times*

"*The Mercury 13* is a book for women young and old, as well as men young and old, who want to learn something that until now has been nearly buried in post–World War II American history."

—*The Denver Post*

"Fascinating . . . Besides being an excellent volume in the category of women's studies, *The Mercury 13* also serves to fill a critical gap in the history of NASA and (wo)manned space flight."

—*BookPage*

"This exciting story is well worth reading. . . . Here were thirteen women who wanted opportunity above all, thirteen women who would have laid their lives on the line to prove that women deserved a chance to become the best they could be."

—*The Women's Review of Books*

"Lovingly researched and beautifully written . . . Ackmann goes at it with gusto." —*The Orlando Sentinel*

"A lively account of the little-known attempt to qualify women as astronauts in the early sixties . . . a work of nonfiction as ambitious as its subjects." —*The Christian Science Monitor*

"*The Mercury 13* ought to be on the shelf next to *The Right Stuff* as a glaring and embarrassing counterpoint to the triumph of the boys' club and its fighter jocks. Ackmann has done outstanding investigative reporting." —WILLIAM E. BURROWS,
author of *This New Ocean:*
The Story of the First Space Age

"[*The Mercury 13*] demonstrates the determination of the women who tried to become astronauts, as well as identifying a few well-placed men who wanted to break the gender barrier. . . . Ackmann juggles a lot of characters while carrying a narrative thread successfully. Her research is impressive."

—*Fort Worth Star-Telegram*

"A fine read that inspires fist-shaking indignation."
—*Wired*

"In dynamic prose, Ackmann . . . relates the story of thirteen female pilots who fought to become part of the nation's space program at its inception. Their tale is uplifting, a narrative of their dedication . . . and sacrifice in an attempt to aid the nation in the space race against the Soviets and to experience the thrill of space flight. . . . An utterly compelling book."

—*Publishers Weekly*
(starred review)

"Impressive research, vivid biographical details . . . and even-handed narration . . . *Mercury 13* [is] an inspirational story about perseverance, courage, and ultimate triumph."

—*Bookmarks*

"Ackmann . . . draws vivid portraits. . . . She has an inherently interesting story, and she tells it well." —*The Atlanta Journal-Constitution*

"Ackmann's account is well-researched, well-written and revelatory. . . . She explores the myths that caused so many otherwise intelligent politicians, space engineers, military officers and even women to reject women as serious astronaut candidates."

—*Austin American-Statesman*

The Mercury 13

Random House Trade Paperbacks | *New York*

The Mercury 13

THE TRUE STORY OF

THIRTEEN WOMEN AND THE

DREAM OF SPACE FLIGHT

MARTHA ACKMANN

LIBRARY OF CONGRESS CATALOGING-IN-PUBLICATION DATA
Ackmann, Martha.
The Mercury 13 : the true story of thirteen
women and the dream of space flight / Martha Ackmann.
p.   cm.
Includes bibliographical references and index.
ISBN 978-0-375-75893-5
1. Women astronauts—United States—History.
2. Project Mercury (U.S.)—History. I. Title.
TL789.85.A1A28   2003
629.45'0082'0973—dc21   2002037118

Random House website address: www.atrandom.com

Printed in the United States of America

Book design by Barbara M. Bachman

For my parents,
Florenze and Elizabeth Ackmann

FOREWORD

BY LYNN SHERR

AT PRECISELY 7:33 A.M. ON JUNE 18, 1983, A BRIGHT FLORIDA MORNING with a radiant sun, the space shuttle *Challenger* was lifted off the launchpad on a white-hot surge of rocket power. Officially, it was STS-7, the seventh trip for the nation's two-year-old space shuttle system, itself a pioneering program. But for most of the folks who were watching that day, *Challenger*'s mission represented something far more revolutionary: the first flight of an American woman into space. Sally K. Ride carried the dreams of her earthbound sisters with grace and good humor. As the shuttle shot skyward with a force far beyond that of any amusement park ride, she immediately connected to Americans everywhere by radioing her counterpart back in Houston: "Have you ever been to Disneyland? This is definitely an E ticket"—in other words, the hottest ride of all.

As I anchored the launch for ABC News that morning, I, too, felt a special sense of pride. It wasn't just that one of us was up there, or that there were five more women astronauts in the wings. It was also my satisfaction that both the spirit and the substance of America's manned space program had changed irrevocably. NASA had finally gotten it right.

Later, I felt even better. Eager to share the excitement, I had invited my sister and my mother to the Kennedy Space Center to see the launch, and when we finally hooked up after the day's news had settled down, they were glowing. "Fabulous," said my sister excitedly, having seen such things only on television up until then. My mother, who was about to be eighty, put it all together: "I've seen the horse and buggy. I've seen the car and the train and the airplane. And now this. Perfect."

It was all perfect—the mission, the leap into equality, the giddy sense of accomplishment for the entire nation. As for the star of the show—the very private but very cool astronaut who had become a close friend—Sally told me in an interview before she left, "I do feel under some pressure not to mess up." She did not. In fact, Sally became the most famous person in the world for the next few weeks, as little girls in droves decided to become astronauts.

Today, three decades later, thirty-six more women have flown on American shuttles, doing virtually everything their male counterparts have done in the vacuum of space. They have donned special suits for space "walks," they have manipulated the robot arm, and they have launched satellites. Shannon Lucid set an endurance record on the space station, *Mir.* Eileen Collins became the first female shuttle pilot in 1995. Five years later, she slid over to the vaunted left-hand seat and became the shuttle's first female commander. Commander. That's the job that used to belong only to men with "the right stuff," spiritual descendants of Flash Gordon and Captain Kirk who'd challenged gravity as test pilots and defied the odds as fighter pilots. Now women were part of the club. As of this writing, thirty women are in the astronaut corps, nearly one fifth of the total.

A more sobering indication that women are achieving parity is a sad one: four women have given their lives for the program—Judy Resnik and Christa McAuliffe on *Challenger* in 1986, and Kalpana Chawla and Laurel Clark on *Columbia* in 2003. Their deaths were no more and no less heartbreaking than those of their male colleagues, but I was especially touched after the *Challenger* explosion when Anna Fisher, a physician who had joined the program with Resnik and Ride as one of the original NASA women, told me, "I'm really going to miss Judy. I'm going to miss getting to have a reunion when we're all eighty years old."

That she could have even imagined a reunion illustrates the point about women in space today: we take it for granted that women will fly, and perhaps die, in space. And it didn't used to be that way at all.

Before I started covering NASA in 1980, there were few female reporters, fewer female NASA managers, and no female American astronauts. Space, like almost everything else in American society, was a men's club—run by, for, and with men. The arguments that preserved that exclusivity were the same tired excuses that once kept us from practicing medicine, running corporations, playing tennis for big money, and casting a ballot. Women were too weak, too emotional, too, well, womanly, to participate. And, of course, they simply weren't qualified.

As Martha Ackmann points out definitively in the pages that follow, that judgment was categorically wrong in this case, as in so many others. The Mercury 13 women were not only dedicated and determined, they were enormously talented. To read about their accomplishments and to discover their passion to fly is to recognize an incredible lost opportunity by the United States government. And they weren't anomalies. There were plenty of qualified females out there, if anyone had taken the time to find and trust them. At the same time, it is both amusing and depressingly familiar to learn that these gifted pilots, charting the skies and scrounging their planes, had to keep their noses powdered and their hemlines straight to impress the world. Some things never change.

Equally astounding is to learn about the men. Not only that so many were so derisive—some might say threatened—at the thought of mere women on the frontier, but that so many took up the cause with such energy. By identifying these supporters and chronicling their activities, Ackmann has given us a new pantheon of heroes.

Every time we turn a page in social history, someone comments that the barriers have fallen for eternity, there's no turning back and that women (or whoever) have won, and the battle is over. Maybe. Whatever happens to the U.S. space program—and I suspect there are some vast changes coming up— I think the presence of women is assured. But they're still a minority. And anyway, I think there are always new envelopes to push.

For example, on the wall of my office is a framed certificate, fading now after the glow of too many sunsets. It's my official recognition as a semifinalist in NASA's Journalist-in-Space program, a short-lived effort with the very sensible goal of putting one of us into the shuttle. I was one of forty who made it to that stage—just before the explosion of the *Challenger* abruptly ended the competition. I still think a reporter should go into space, and I'd still volunteer. If not me, another woman. Or even a man. We can afford to be generous when the doors finally open.

CONTENTS

THE THIRTEEN AMERICAN WOMEN PILOTS WHO PASSED THE PHYSICAL tests for astronauts at the legendary Lovelace Foundation never identified themselves by a group name while they were undergoing the exams in 1960 and 1961. Dr. W. Randolph Lovelace II, who earlier had administered the same physical tests to the Mercury 7 male astronauts, did not assign a project name to the women's secret testing program. The first woman to undergo the tests, Jerrie Cobb, later referred to the women as F.L.A.T.s (Fellow Lady Astronaut Trainees), but the other women disliked the name, and it did not stick. The fact that the group never took on a collective name contributed to its historical invisibility. More than thirty years after their testing, media attention—triggered largely by John Glenn's second launch into space—revived interest in the thirteen women and the role they played in the history of U.S. space flight. In editorials, a television documentary, NASA resources, library archive references, and citations from professional organizations such as Women in Aviation, International, the women began to be called "the Mercury 13." It is the name most frequently used now to refer to the women and the one most of them prefer.

The Mercury 13

CHAPTER ONE

Space Fever

JERRIE COBB REACHED DOWN AND PULLED THE HEAVY LAYERS OF arctic clothing over her navy blue linen dress. Already the temperature on the airport tarmac that afternoon in June 1957 was a steamy ninety degrees. The shy, soft-spoken young pilot did not mind the heat nearly as much as she minded the reporters who crowded around her. She disliked all the attention and being forced to answer questions such as why she needed warm clothing for her attempt at a new altitude record. Cobb had trouble putting her thoughts into words and knew reporters found her not as quotable as they would like. The questions were predictable. "Are you frightened, Miss Cobb, about trying to break the world record?" "How cold will it get up there?" "Why does a pretty young girl like you want to spend her time around the dirt and grime and noise of airplanes?" "What about boyfriends? Are you more afraid of dating than flying six miles up?" Cobb paused before answering and patiently tried to explain why flying was more important than anything else in her life. It was always difficult for her to describe how content she felt when she was alone in an airplane. She realized her words sounded flat and could never express the genuine passion she felt for flying. It was easier to keep her personal feelings hidden and focus instead on what she wanted to accomplish that day. Her goal, Cobb told the reporters, was to the break the current world altitude record for lightweight aircraft. Since Oklahoma was celebrating its semicentennial, she wanted to use her skills as a pilot to set world records for the Sooner State. Aero Design and Engineering, an Oklahoma City–based aviation company, had been eager to sponsor Cobb and lend her its new twin-engine airplane for the record-breaking flight. It was good publicity, especially after Cobb used one of their planes to break the world record for nonstop long-distance flying—from Guatemala City to Oklahoma City—just five weeks earlier. Today Cobb would push the Aero Commander beyond the highest altitude it had ever achieved. It was a risky proposition. Test pilots had flown the plane to 27,000 feet. Cobb was hoping for 30,000. Her parents just hoped that she could avoid a fatal stall.

Cobb excused herself from the clutch of reporters to concentrate on her final checklist. She had been up since daylight to smoke the barograph drums. Taking a stick of camphor, Cobb had held it near the barograph, coating the surface with dusky smoke. The sharp point of a stylus would scratch through the soot to register her precise altitude. To prepare her lungs for the thin air of the upper atmosphere and wash out nitrogen in her system, she breathed 100 percent oxygen for two hours before the flight. She walked around the aircraft, examining the plane's hinges, vertical stabilizer, and rudder, and got down on her knees to check the tire pressure. She lowered the plastic fuel sampler and studied the color of the fuel, checked it for sediment, and smelled its sharp, pungent odor. Standing nearby, Cobb watched as officials from the National Aeronautic Association certified the official scales and confirmed the Commander's weight class. Cobb kept any concerns about the dangers of high-altitude flying to herself. She knew that at several miles up and at high accelerations, breathing became difficult, vision was impaired, and a pilot could faint. She had been told about the terrifying slump toward unconsciousness: first color would disappear from one's vision, turning everything gray, then sight would shrink to a narrow tunnel, and, finally, all would go dark. Cobb knew she would be needing oxygen bottles in the unpressurized cockpit above 12,500 feet in order to maintain consciousness. She also knew that—as absurd as it seemed—she had to worry about her appearance as well. Unspoken social customs for women pilots dictated that she wear a dress and high heels under her protective clothing. Everyone expected women pilots to look like fashion models when they stepped out of a cockpit, even if they had been up all night working on engines with their arms covered in grease. When she realized she had forgotten a mirror and would need one for retouching her makeup before she landed, Cobb accepted a small compact from a bystander. Then she climbed into the cockpit and, alone at last, kicked off her high heels.

Cruising to normal altitude, Cobb looked below to the flat Oklahoma prairie and at the sky all around her. She always thought the sky looked bluer when she was actually in it than it did from the ground. She pitied people who spent their entire lives earthbound. They were missing quite a show, she thought. As the Aero Commander rose to 27,000 feet, Cobb could feel the strain on the aircraft and her own body. Every hundred feet meant that both the plane's engine and her breathing were more labored. Cobb breathed from an oxygen bottle and lifted the Commander's nose upward. One more foot, two more feet, she seemed to tell the plane, keeping her eye on the altimeter's needle. The higher she flew, the colder it became in the cockpit. At thirty degrees below zero, the windshield slicked over with ice and the instruments inside became frosted. Flying as much by touch and instinct as by instrument, Cobb continued to push upward almost inch by inch. That was when the beeping started. The stall indicator triggered its alarm. The closer together the beeps were, the nearer Cobb was to a stall. She could either lower the nose to avoid a shutdown, or keep listening to the interval between beeps and praying she had time to climb a few more feet for the world record. Cobb eased the plane upward as images of disaster crowded her mind. If the engine stalled, the Commander would plummet in an unrecoverable spin, faster and faster, wings and tail shearing off in a fatal dive to the ground. Cobb listened to the beeps. Was there just enough time between them to raise the nose once more? Could she gain the remaining altitude for the record? Cobb adjusted the fuel flows and pulled upward. She looked at the altimeter, bouncing higher and higher, 28,000 feet, 29,000 feet, 30,000 feet. At 30,330 feet, the Commander started to shudder, but Cobb almost smiled. She had clinched the world record.

As Cobb guided the plane down, warmer air began to melt the icy windows, and she breathed more easily, as much from relief as from altitude. Floating down from the clouds, she saw the crowd waiting for her at the airport below. Now comes the hard part, she thought. World record behind her, it was time to smooth her blond ponytail, put on lipstick, squeeze her feet into high heels, and start talking.[1]

---

[1] A mechanical problem was later discovered on the barograph measuring device Cobb had been given by aviation authority officials, and she had to repeat the flight. She flew again for the same altitude record in 1960, and the barograph accurately recorded an even higher altitude than she had achieved in 1957—37,010 feet.

Although talking did not come easily to Jerrie Cobb, the press attention greeting her at the airport should have felt routine by now. At twenty-six, she had been making headlines for years. She was the only woman in the United States to have ferried military surplus aircraft to countries in South America, Europe, and Asia. She could tell dramatic stories of flying solo over the jagged Andes, hopping a ride on banana boats after emergency landings, and sleeping in less-than-hospitable surroundings with her pistol nearby. An air-race competitor in women's cross-country and international derbies and now a world record holder, Cobb was reaching the goals that every topflight pilot, man or woman, wanted to attain. More than anything, a great pilot wanted to go "higher, faster, and farther"—the four words were considered a champion's credo. Some people joked that Cobb, with her two new world records, was becoming the country's best Cold War weapon. When she entered the record books the previous month by breaking the nonstop long-distance mark, people in Oklahoma City pointed out that a local girl in a local plane beat a world record held by a male Russian pilot flying a Soviet Yak II aircraft. Jerrie Cobb seemed to be single-handedly winning aviation contests against the Russians—"Sooner-Soviet air competitions" one newspaper reporter called them.[2]

Cobb had gone both higher and farther and had the world records to prove it. There was one record left to complete her aviation hat trick—the world record for speed. Setting off before a cheering crowd of 7,500 at the Las Vegas World Congress of Flight in 1959, Cobb raced the clock over Reno, San Francisco, and San Diego and back to Las Vegas in another twin-engine Aero Commander. Representatives from the National Aeronautic Association and the Fédération Aéronautique Internationale of Paris, the official authority on aviation world records, timed her flight and took two sealed boxes from her plane after she landed for shipment to the U.S. Bureau of Standards for speed verification. When the results came back, Cobb had secured a third world record, surpassing another male Russian pilot as the holder of the light plane speed record.[3]

Cobb had already been named the 1959 "Woman of the Year in Aviation" by the Women's National Aeronautical Association and "Pilot of the Year" by the National Pilots Association. When she came home after her third world record, the chief of flight operations at Oklahoma's Tinker Air Force Base said that Jerrie Cobb's records beating the Russians ranked in importance to military victories. The Oklahoma City Chamber of Commerce president agreed. "Miss Cobb," he said, "seems to be taking records away from the Russians. Maybe she could help President Eisenhower."[4]

Escalating Cold War hostilities and Dwight Eisenhower's apparent non-chalance about Soviet achievement in space were very much on the mind of the American public. Just months after Cobb set the altitude record, the unearthly beep of Russia's first satellite, *Sputnik*, shocked a sleepy nation, shook it awake, and forever linked the military objectives of the Cold War with space exploration. Within hours after its launch, the satellite's beep was picked up by shortwave radio operators and recorded for broadcast over television and radio stations in the United States. Listeners found the sound, so far away from Earth, both thrilling and terrifying. If Communists were moving into outer space, they could dominate the rest of the world as well. It seemed as though President Eisenhower was the only American who did not initially understand the military, scientific, or cultural significance of *Sputnik*'s chirp.* "The Russians have only put one small ball in the air," he said.[5] Others disagreed. The Democratic governor of Michigan, G. Mennan Williams, went so far as to compose a poem to Eisenhower's golf-playing detachment:

> *Oh little Sputnik, flying high*
> *With made-in-Moscow beep,*
> *You tell the world it's a Commie sky*
> *And Uncle Sam's asleep.*
>
> *You say on fairway and on rough*
> *The Kremlin knows it all,*
> *We hope our golfer knows enough*
> *To get us on the ball.*[6]

Sputnik's launch prompted the United States to reevaluate its presumed world superiority in education, industry, and defense. Teachers reexamined

---

*Paul Dickson, in his recent book, *Sputnik: The Shock of the Century*, argues that Dwight Eisenhower was more aware of the consequences of *Sputnik* than was first believed. Dickson contends that President Eisenhower believed *Sputnik* justified the concept of "freedom of space," opening the way for United States surveillance satellites to observe Soviet missile action.

what U.S. schoolchildren were learning and were appalled to discover how far behind the nation's students were in mathematics and science. American scientists studied the gap between Soviet and U.S. technological achievements and were equally frustrated to find their accomplishments lagging. Record-setting pilots such as Jerrie Cobb realized that "higher, faster, and farther" now meant something totally different since a satellite had been launched. Flying in outer space became the new frontier for top-flight pilots. But beating the Russians in spaceships was nothing like beating them in airplanes. Cobb realized that an individual pilot's determination and skill were not enough to outpace the Soviets in space. That competition would take a national effort, millions of dollars, thousands of hands, and one sophisticated spacecraft.

Even President Eisenhower had to agree that the United States was in second place in space. In an effort to address the national concern, he signed into law in 1958 the National Aeronautics and Space Act, creating the National Aeronautics and Space Administration (NASA). Although the U.S. Air Force, Army, and Navy had all jockeyed for control of the space agency, Eisenhower decided NASA would be under civilian rather than military control and gave the new federal agency a broad mandate to challenge the Soviets in aerospace superiority. With a high-profile charge, the young federal space agency faced myriad challenges, including fundamental questions about what its top priority should be. Scientists within the organization argued that NASA should emphasize the acquisition of scientific and technical knowledge. Spaceflight, they said, should not be focused on meaningless races in which finishing first was the only goal. It did not matter to them if the United States launched the first satellite or even the first living creature—man, woman, dog, or chimpanzee. Such competitions were better suited for athletic fields and obscured the more important scientific objectives of spaceflight. Others, however, both within and outside NASA, saw space exploration as connected to larger national and even international objectives. Successfully launching a spacecraft vividly demonstrated a country's accomplishment and brought it worldwide prestige. Space launches were dramatic, captivating the public's attention with their suspense and spectacle. Thunderous noise, bright fists of flame, sleek rockets rising higher and higher into the sky—the images were tailor-made for the new medium of television. Viewers felt a personal connection and a sense of national pride when rockets lifted off from what seemed like their very living rooms.

Indeed, for the general public, the idea of beating the Russians by launching a man into space seized their attention in ways that sending up an unmanned missile for scientific purposes did not. But who would be that first man in space? Eisenhower initially believed that astronauts should come from a variety of professions—arctic explorers, mountain climbers, meteorologists, flight surgeons, deep-sea divers. People with a wide range of abilities and perspectives would enhance space exploration, he thought. But the President changed his mind. In late 1958 he decided that NASA should narrow the field and choose astronauts from the ranks of military jet test pilots, a field that barred women and included few minority men. The shift in Eisenhower's thinking reflected the urgency he felt for launching a U.S. manned space program before the Soviets had a chance. His decision also was informed by his experience, respect for military protocol, and the advice of NASA officials. Dr. T. Keith Glennan, NASA's first administrator, argued that his own years in the service had convinced him that military jet test pilots would make the best astronaut candidates. Men who flew for the military were already admired for their skill, experience, and courage, he said. Why open up the selection process to anyone else when it would be more efficient to survey a smaller, recognizable group instead? Glennan presented his case to the President and "got it cleared in five minutes," he later recalled.[7] The quick decision made in the White House that day set into motion a series of extraordinary events that would soon make history. Eisenhower's verdict also perpetuated another kind of history, a legacy of exclusion. In opting for the fast and the familiar, the President placed expedience ahead of equity and did not adequately consider whether his judgment compromised fundamental principles of democracy. It never occurred to Eisenhower or Glennan that someone other than white men might also have the desire and the ability to fly in space.

When the call for astronaut candidates went out to the military, not every jet test pilot was interested. Some pilots at Edwards Air Force Base, for example, saw themselves as doing more important work than the astronauts would be required to do. Sitting in a space capsule was not flying, they said. It was riding. As test pilots they made life-and-death decisions daily, flying by instinct and experience. They certainly did not want engineers calling the shots for them. Some pilots called the future astronauts "guinea pigs," "Spam in a can," nothing more than "subjects" inserted into a capsule and told by men with slide rules not to touch anything. Even chimpanzees were being discussed as NASA's first live cargo. Many hotshot test pilots did not want to be equated with a chimp, no matter how exciting the launch might be. It did not help matters that the man who had broken the sound barrier, the most celebrated American pilot of all, Chuck Yeager, mocked the astronauts by saying that before they could fly, they would have to sweep monkey shit off the seat of the space capsule.[8]

In April 1959, the American public got its first look at the seven young men who would propel the country into the manned space race. "The nation's Mercury astronauts," Keith Glennan announced with great fanfare, as the NASA press conference erupted in applause and photographers rushed to the front of the room for closeups.[9] Malcolm Scott Carpenter; Leroy G. Cooper, Jr.; John H. Glenn, Jr.; Virgil I. "Gus" Grissom; Walter M. Schirra, Jr.; Alan B. Shepard, Jr.; and Donald K. "Deke" Slayton—they were as all-American as a John Philip Sousa march, smiling, crew-cut, stand-up-and-salute soldiers from small towns across the country. Between the ages of thirty-two and thirty-seven, they were the shining embodiments of their middle-class, white, Protestant backgrounds. Reporters immediately wanted to know everything about them: why did they want to go into space; what did their wives think of their dangerous work; how did their religious beliefs square with the idea of spaceflight; who would be the first one to be launched? Also seated on the dais that day were two other men, just as eager to talk about the Mercury program as the seven astronauts. Dr. W. Randolph Lovelace II, chairman of NASA's Life Sciences Committee, and Brigadier General Donald Flickinger of the Air Force helped design the medical testing procedures for the astronaut candidates that took place at the Lovelace Foundation in Albuquerque and at the Wright Air Development Center's (WADC) Aeromedical Laboratory at Wright-Patterson Air Force Base in Dayton. They played a central role in selecting the Mercury 7. The fifty-one-year-old Lovelace, an engaging and well-respected man, knew his astronaut tests had been formidable. "I just hope they never give me a physical examination," he joked, glancing at the astronauts with a wink. "It's been a rough, long period that they have been through. I can tell you that you pick highly intelligent, highly motivated and intelligent men, and every one is that type of person. . . . I can tell you that I am very, very thrilled that we have had a part in the program."[10]

What Dr. Lovelace did not say at that moment was that he and Flickinger were interested in testing women for potential spaceflight and had a hunch that females might offer some advantages over males as astronauts. They wondered if women's lower body weight, for example, would make them better human cargo for American rockets, which were having a difficult time lifting heavier payloads. Every pound in the spacecraft necessitated more booster power. Greater human weight also required a greater oxygen supply and more food. If the weight of the space capsule could be reduced even slightly by using a lighter astronaut, the ability to boost the capsule would be less of a concern. Lovelace and Flickinger were also curious to determine whether women could measure up to the same demanding physical standard that the Mercury astronauts established. Were women physically weaker, less resilient, less capable of dealing with isolation, stress, and danger? Lovelace and Flickinger were reluctant to accept general assumptions about women's inferiority and wanted to scientifically evaluate women and compare their data to the men's. As scientists intimately connected to the international community, they also knew that the Soviets were already discussing the possibility of women astronauts. Some rumors coming out of Moscow suggested that Russia might even launch a woman in their first orbital spaceflight, and everyone believed the first human mission was imminent. At the very least, Lovelace and Flickinger wanted to test one woman on the same medical trials the seven Mercury astronauts just had completed under their supervision at Wright-Patterson and the Lovelace Foundation. If a single woman test subject did well, perhaps others would also. They just needed to find the right volunteer. For them, an exceptional woman pilot would make the ideal candidate.

Within the months following the astronaut press conference, the seven Project Mercury astronauts became national celebrities. *Life* magazine began telling their personal stories in glossy photo essays about their families and rigorous NASA training. They were young, energetic, skilled, and attractive and fit the prevailing hero mold in every way, including their appetites for anything fast. Although the public looked upon some of the astronauts as cocky—even a little reckless—with their Corvettes, Austin Healys, and rumors about extramarital affairs, they seemed to excuse these foibles as being those of red-blooded American boys. The astronauts' wives became well known, too, and their images were required to fit another, equally familiar mold. *Life* featured the women in staged photographs listening to their husbands' stories of the space capsule training, keeping their husbands company as they did their astronaut homework, or being tutored in golf so that they could play with their husbands. The message the photographs conveyed was that women always watched, waited, helped, and learned from men. A *Life* caption perfectly captured the passive role society expected women to play when it described three astronaut wives as trying out facials "to pass the time."[11]

When voters elected John F. Kennedy in 1960, many looked to the new chief executive as another vigorous man of action who could propel the country beyond the Russians, making the United States preeminent in the world and in space. Kennedy knew he had to clean house at NASA, bring in his own administrator, and set new priorities. Already Eisenhower's NASA chief, T. Keith Glennan, had announced his resignation, exhausted from constantly reassuring the public that the United States was doing its best to beat the Soviet Union. The criticism over NASA's ineptitude when it came to launching rockets had skewed into ridicule. So many rockets ended up exploding on the launch pad that they earned sarcastic nicknames. The Navajo became the "Never Go." Snark missiles that frequently ended up in the waves off the Atlantic caused "Snark-infested waters." The old Cape Canaveral joke seemed sickeningly apt. "You know how children at Cape Canaveral learn to count, don't you?" the joke went. "Five, four, three, two, one, damn it!" Even the astronauts themselves had reasons to doubt NASA's ability. John Glenn never forgot the spring night in 1959 when NASA brought the newly selected astronauts to the cape to witness an Atlas launch for the first time. The seven men stared at the gantry holding the missile as the countdown reached zero, the engines ignited, and the rocket rose and then exploded. They stood still as debris rained over the ocean, soberly aware of the personal risks they soon would be taking. Not a word was spoken until Alan Shepard attempted some dark humor. "Well, I'm glad they got that out of the way," he said.[12]

Kennedy knew that the jokes about NASA's incompetence had to stop, and he needed to find an energetic leader for the space agency who could turn things around and give the country reason to believe the United States could beat the Russians in outer space. He also needed a man who could negotiate the festering administrative problems that consumed NASA—battles among scientists, politicians, and the military about who would control the space agency and what missions should come first. But Kennedy's staff could not find a man who was right for the job or who was willing to take it. They had crossed off a dozen men from the list of possibilities before the head of the Senate's space committee suggested James Webb of Oklahoma.[13] Known as an efficient administrator, Webb had served as President Harry Truman's director of the Bureau of the Budget and later worked as undersecretary of state. When he left Washington and moved west to work in the oil industry, Webb plunged into a variety of statewide efforts to promote science education and bring aerospace business to Oklahoma. He urged leaders of the state to look to the future and develop Oklahoma by emphasizing aviation and space rather than clinging to its cowboy past.[14] Kennedy liked what he heard.

On a snowy January afternoon, a few days after John Kennedy had been sworn in as President, Jerrie Cobb sat down next to James Webb for a Chamber of Commerce luncheon in Oklahoma City honoring the state's leaders in the aerospace industry.[15] One could not tell by looking at her quiet demeanor, but Cobb was sitting on secrets. After a chance meeting with Dr. Lovelace and General Flickinger just months after the press conference introducing the Mercury 7 astronauts, Cobb became a test subject at the Lovelace Foundation. Lovelace and Flickinger had found their exceptional female pilot. When Lovelace later revealed the astonishing results of Cobb's test scores at a scientific conference in Stockholm, reporters began calling Cobb's parents in the middle of the night, trying to track down the taciturn young woman the media instantly dubbed "America's first woman astronaut." The past year had been a whirlwind for Cobb and now she held another confidence. Dr. Lovelace had just begun to test twelve more women pilots and she had helped select them. The first candidate had already arrived in Albuquerque at the Lovelace Foundation and more women would soon follow that spring and summer. All the women had been pledged to secrecy, but Cobb knew their identities and she considered them exceptional pilots and ideal test subjects. There were identical twins from California, Jan and Marion Dietrich; the youngster, twenty-two-year-old Mary Wallace "Wally"

Funk from Taos, New Mexico; the owner of a flight operation in Michigan, Bernice "B" Steadman; the Air Force Reserves officer from Akron, Jean Hixson; the flight instructor from Georgia, Myrtle Cagle; the engineer from Kansas City, Sarah Gorelick; the executive pilot from Houston, Rhea Hurrle; the forest service pilot from Chicago, Irene Leverton; the aviation instructor from the University of Oklahoma, Gene Nora Stumbough; the sassy air-race competitor from Dallas, Geraldine "Jerri" Sloan; and the U.S. senator's wife from Michigan, Jane "Janey" Hart.[16]

As Jerrie Cobb made idle conversation with James Webb at the luncheon's head table, she hoped he wouldn't ask her about the New Mexico tests. The day before, *The New York Times* had printed a brief article reporting that twelve unnamed women were undergoing astronaut exams. Cobb did not want to be asked about the women's identities or any other details that would wind up in another newspaper.[17] But Webb did not have as much time to talk with Cobb as he expected. While luncheon dishes were being cleared, he received a phone call from the science adviser to President Kennedy. Webb left the dais, walked out of the room, and took the call. What he heard surprised him. Kennedy wanted him in Washington immediately to discuss with Vice President Lyndon Johnson the prospect of taking over leadership of NASA. Though flattered, Webb was cautious, thinking that a scientist would be better suited for the job. The White House persisted and Webb finally agreed to fly out to Washington that evening. Webb hung up the phone and returned to his seat next to Jerrie Cobb. Staring at the desserts in front of them, both Cobb and Webb tried not to divulge their space secrets.[18]

After meeting with Johnson and Kennedy, James Webb surprised himself by reversing his initial impulse about the job. "I never said no to any President," he said.[19] Within moments of Webb's acceptance, Kennedy shuttled Webb to his press secretary and then disappeared to prepare his first State of the Union address to be delivered that evening. Standing before the White House press corps, perhaps a little stunned by the sheer speed of the decision, Webb greeted reporters as President Kennedy's nominee for NASA administrator. As soon as he was able to get to a phone, Webb called his wife with the news. The radio was on, she told him, and she had just heard the announcement.[20]

That evening, with only ten days as President behind him, Kennedy delivered his State of the Union address. It was the young president's opportunity to separate himself from the previous administration, to take bold strides and claim new ground. What he had learned in the last week and a half was "staggering," the President admitted, and there were reasons for the nation to be both optimistic and concerned. The economy, defense, even the exploration of outer space needed aggressive action, he declared. While the United States was leading the Soviets in scientific research of the solar system, Kennedy said, the Russians were ahead in building powerful rockets capable of hoisting a man into orbit. To those who were listening carefully, Kennedy's news about the space program did not arouse much enthusiasm. Leading the world in scientific research was not half as thrilling as being poised to launch a man into orbit. It was like saying the United States was ahead in chalkboard equations while the Soviets were practically counting down to blastoff.[21]

Russia would launch a manned space capsule even sooner than Kennedy imagined. Telephones started ringing on Capitol Hill at 3 A.M. on April 12, 1961, as word came from Russia that Major Yuri Gagarin had orbited the earth. When reporters wanted a response from NASA, they called the "Voice of Mercury Control," John "Shorty" Powers, and found that he was still in bed. "We're all asleep down here," Powers confessed.[22] Brigadier General Flickinger was in the midst of a press briefing at Cape Canaveral when he was handed a note. "They've got a man in orbit," he quietly told the group.[23] Within a matter of hours, President Kennedy's honeymoon with Congress came to an abrupt halt. People wanted answers. Pennsylvania Representative James G. Fulton argued that the United States simply was not working hard enough to send a man into space. "I am getting awfully tired of the Mother Hubbard approach of 'Tie your apron up after the Russians do it.' All you have to do is put a little overtime on and go around the clock on some of these programs instead of knocking off at 5 o'clock. I think we in the United States should stand the expense of it and put some overtime in on this and pay for it."[24] Fulton's exasperation spoke for many Americans: it was time to dig in and get to work.

President Kennedy held a news conference a few hours later and reporters immediately asked if the United States could catch up to the Russians, much less surpass them in the space race. He curled his hand into a fist and thumped the podium twice, making sure everyone—perhaps no one more than himself—understood the reality of the situation. "The news may be worse before it gets better, and it will be some time before we catch up. . . . We are, I hope, going to go in other areas where we can be first and which will bring perhaps more long-range benefits to mankind. But here we are behind." The President fidgeted with the papers in front of him, stammered a bit, and concluded, "I do not regard the first man in space as a sign of the weakening of the free world."[25]

That evening CBS interrupted its regular broadcast of *Malibu Run* to present an hourlong special on the day's events. Opening the program, the announcer declared that the Soviet feat would be recorded as one of mankind's greatest achievements. Babies born in the Soviet Union today are named Yuri, he said. In Moscow, if your last name is Gagarin, everyone wants to buy you a drink because all Soviets are celebrating. Russian leader Nikita Khrushchev taunted the West by saying, "Now let the capitalist countries try to catch up." In interview after interview with American scientists, politicians, and NASA officials, the tone was defensive and the optimism unconvincing. James Webb looked like a man with his back against a wall. Huddled in an overcoat against the early spring chill, Webb tried to minimize the damage to American pride: "What happened last night is the result of many years of work. We are expecting before the end of the year to have an orbital flight." A senator standing with Webb for the interview could not raise his eyes and stated flatly that the country was "nine months to a year behind." One reporter asked if the Mercury program should be discontinued now that the Russians had launched a man first. That would be like "saying Chevrolet should never have started making automobiles because Henry Ford got there first," astronaut John Glenn protested. Yet no matter how much anyone rationalized Gagarin's orbit, the United States had failed to launch the first man into space. America was losing to the Soviets.[26]

John Kennedy would have less than a week to focus his attention on NASA as international events intervened. Five days after Gagarin's launch, an armed force of Cuban exiles landed on the south coast of Cuba at the Bay of Pigs. Trained by the CIA and carrying arms supplied by the U.S. government, more than a thousand rebels stormed onshore to overthrow the Communist regime and depose Fidel Castro. The Cuban Army swiftly defeated the insurgents, killing nearly a hundred rebels and leading the rest out of the swamp with hands above their heads in surrender. Protestors staged anti-American demonstrations in Europe and South America. Critics railed against President Kennedy for what they saw as a poorly planned invasion with inadequate air support. Others condemned Kennedy for approving the mission in the first place. Such a humiliating debacle coming so soon in the new administration's term was a serious blow to Kennedy and his staff. They needed to turn the mood of the country around so that the public would have faith both in the President and the country. Defeated, frustrated, angry, and embarrassed, Kennedy sent a memo to Vice President Johnson the next day asking him if NASA was in a position to claim a significant achievement in space. He wanted to know "Are we working 24 hours a day on existing programs? If not, why not? Is there any other space program which promises dramatic results in which we could win?"[27]

But before the United States could win a space race, it first had to catch up. On May 5, 1961, at 9:34 in the morning, Alan Shepard lifted off from Pad 5, becoming the first American man in space. The red glare from *Freedom 7*'s booster rockets lit up a bank of windows in the concrete blockhouse where engineers sat in tense anticipation. As the ground rumbled beneath them, Project Mercury engineers heaved a collective sigh of relief. At least the rocket was strong enough to lift Shepard off the ground.* Even though Shepard's flight was suborbital and only fifteen minutes long, it did mean that the United States had finally reached outer space. A relieved James Webb gave the speech he wanted to make that day, a speech commending NASA on a successful launch. He did not have to use one of two other speeches he had prepared: one telling the world that Shepard had

---

*One engineer who had hoped to be in the blockhouse that morning was Joan Fencl Bowski of McDonnell Aircraft in St. Louis, the aviation company that designed the Mercury space capsule. With her NASA clearance in hand and her immediate boss's approval for transfer, Bowski received word from McDonnell Vice President Walter Burke that she could not go. Although neither Bowski nor her male colleagues ever protested the action (coworkers told her a conversation with Burke would be ill advised), Bowski learned that Burke had refused her transfer because there were no bathroom facilities for women at the NASA site. "It was so stupid," Bowksi said. "I was mad and discouraged and completely put out." "This was the first manned space capsule and I wanted to be part of it all the way. I loved the work." When Shepard's launch went up, Bowksi was back in St. Louis working on another McDonnell assignment. Although disappointed not to be part of the excitement at Cape Canaveral, Bowski knew her contribution to the historic effort could not denied, although it probably would be forgotten. "That work on the escape rocket was mine," she said with pride.[28]

ejected due to a malfunction, or even worse, one that announced that Shepard had been killed.[29]

Jerrie Cobb hoped to attend the Shepard launch but was in New York for a talk before the Aviation/Space Writers Association. Following the *New York Times* article, other reports were leaking out to the public about the testing of women pilots in Albuquerque, and Cobb had begun speaking about her own astronaut examinations, which had taken place a little over a year before. She used her speech to outline the weight advantages of women astronauts compared to men and described the women's testing program currently under way in New Mexico. "These women have volunteered their services," Cobb said, "and have been trained and tested for future astronaut roles at no expense to the Government." Cobb was careful not to discredit the courage or skill of the current corps of male astronauts, but she believed it was important to point out that women were not permitted to be astronauts because females were prohibited from being military jet test pilots. "I see no reason why civilian pilots should not be used," she argued, "thereby releasing military test pilots to carry out their important duties in our defense system." Cobb was becoming more outspoken about her desire to become an astronaut and ended her speech with a clear desire to join Shepard in the space race. "Russia may have put the first man in orbit," she said, "but the United States can now put the first woman in space—and here's one who'd like to be riding that Redstone tomorrow."[30]

At a news conference later that day, a triumphant President Kennedy said Shepard's mission proved that the country "should redouble efforts" in space.[31] In the Vice President's office, Lyndon Johnson hurried to finish work before leaving on a two-week fact-finding trip to Southeast Asia, where he was to assess the need for future American aid to Saigon. Several tasks remained on Johnson's desk before he departed for Vietnam, namely delivering an answer to the President's question about the nation's efforts in space. Johnson sent a directive to NASA and the Department of Defense and asked them to submit to him within three days their reports on the nation's space priorities.[32] While it was gratifying and a relief to finally have a man in space, Johnson needed to know if NASA could surpass the Russians anytime soon.

The public response to Shepard's launch was overwhelming. People lined the streets of Pennsylvania Avenue to get a glimpse of Shepard and the other six astronauts as they rode in open convertibles on their way to the Capitol a few days later. Kennedy took keen note of the nation's sentiment. With racial battles being fought in Alabama between Freedom Riders and segregationists and a political crisis boiling in Laos, Kennedy knew that his administration and the country needed to be revived and unified behind a momentous goal.

That weekend James Webb and his NASA staff worked until 3 A.M. on the space priorities proposal and then passed the secret memo to Defense Secretary Robert McNamara for his signature and delivery to Vice President Johnson. "It is man, not merely machines, in space that captures the imagination of the world," the proposal read. "Dramatic achievements in space therefore symbolize the technological power and organizing capacity of a nation. . . . Our attainments are a major element in the international competition between the Soviet system and our own."[33] The memo went on to detail objectives for Project Apollo and landing a man on the moon. On the afternoon of the Shepard motorcade through Washington, Kennedy made up his mind.[34] He had found a space race worth winning.

On May 25, before a packed joint session of Congress, John Kennedy outlined what he saw as "urgent national needs." The President flashed a broad, confident smile and began. "These are extraordinary times and we face extraordinary challenges," he said. McNamara sat next to Attorney General Robert Kennedy in the front row and listened as the President called for more defense against a possible Soviet nuclear attack. Fallout shelters, the President said, "would protect people against radioactive material." Then, looking up at the senators and congressmen before him—a sea of 518 men and 19 women—Kennedy spelled out his goals for space. It is "time for a great new American enterprise," he said. The country had never before articulated specific, long-range goals for outer space and must take a leap in full view of the world. In rallying the support of a nation, the President called for "every scientist, every engineer, serviceman, every technician, contractor, and civil servant [to give] his personal pledge that this nation will move forward with the full speed of freedom in the exciting adventure of space." In concluding, Kennedy delivered his bold and historic challenge for the world to hear, "that this nation should commit itself to achieving the goal, before this decade is out, of landing a man on the moon and returning him safely to earth. No single space project in this period will be more impressive to mankind, or more important for the long-range exploration of space; and none will be so difficult or expensive to accomplish."[35] James Webb had received his orders from the President: NASA was going to the moon. Jerrie Cobb and the other women of the Mercury 13 wondered if they would be going anywhere at all.

CHAPTER TWO

Taking a Leap

IT'S NO WONDER JERRIE COBB BECAME A PILOT. SPEND EVEN AN hour in Oklahoma, and you see that everything takes flight. Consider, for example, Highway 35, running north–south between Oklahoma City and Ponca City, where Cobb's parents lived: highway signs shudder back and forth on their metal poles, candy bar wrappers dropped from pickup trucks race along the breakdown lane, and hawks, so still they seem to be paintings, float overhead on oceans of air. Eventually anything in Oklahoma that is not tacked down is lifted up and hurled into the immense western sky.

For nearly all of her childhood, Jerrie Cobb was in motion. Within weeks after her birth in Norman, Oklahoma, she moved with her parents, William Harvey and Helena Stone Cobb, and older sister to Washington, D.C., where grandfather Stone, a Republican, served in the U.S. House of Representatives. Ulysses Stevens Stone earned a lot of money and lost a lot of money in Oklahoma. Working in the oil industry, real estate, banking, even the caramel-popcorn business, Stone was the epitome of Oklahoma boom and bust. When Stone lost a reelection bid, he and the Cobb family returned to Oklahoma, where Harvey Cobb worked as an automobile salesman. A few years later, the Nazi invasion spread across Europe and Harvey Cobb's National Guard unit was activated. The family moved to the Army Air Corps base in Wichita Falls, Texas, and then Denver before finally heading back to Oklahoma after World War II. Helena Stone Cobb was glad to return to her roots. Like her husband, she was a graduate of the University of Oklahoma, and she especially enjoyed keeping up with her Gamma Phi Beta sorority. A dinner guest at the Cobb home remembered Mrs. Cobb's way of identifying women: "Oh, you remember Sally, she's a Kappa Kappa Gamma" or "That's Jane, she's Kappa Alpha Theta." It almost seemed that no matter what a woman might accomplish in life, nothing was more important to Helena Cobb than her sorority affiliation.[1] Certainly her younger daughter felt her mother's judgment. "Mother loved me but never really understood me," Cobb later confessed.[2] Helena Cobb was especially bewildered about all the time father and daughter spent reconditioning an old two-seater Waco plane that Harvey Cobb bought in Wichita Falls.[3] Then again, no matter where they were, Jerrie Cobb seemed to tag along after her father. Cobb's older sister, Carolyn, believed their father would have liked a son. From her point of view, "Jerrie took the place of that."[4]

Being around airfields during her early years was a formative experience for Jerrie Cobb. So, too, was growing up in new environments and frequently among strangers. Cobb coped with being an outsider by adopting two strategies: keeping her mouth shut and staying by herself. When an early speech impediment caused Cobb to slur her words, she felt odd and ridiculed. Later, writing in her autobiography, she called her early problems with speaking "anguish."[5] An indifferent student who preferred the outdoors and physical activity, Cobb took to skipping school for weeks at a time. When Harvey Cobb discovered his daughter's truancy while the family was living in Denver, he put her in the car and drove up to Rocky Mountain National Park for a serious talk. Cobb knew he could reach his daughter more effectively by surrounding her with trees and the silence of nature. Besides, he was better at reasoning with her than his wife was. Harvey Cobb persuaded his daughter that she would never have a future without completing school. She reluctantly returned to class, firmly convinced of three things: school was no good, talking was worse, and the best fun was being alone.[6]

Harvey Cobb was shrewd in appealing to his daughter's plans for the future. Jerrie Cobb knew what she liked and what she wanted to do. She wanted to fly airplanes. After Harvey Cobb got the old Waco running, he took his daughter for a ride. For twelve-year-old Jerrie, flying was like finding her home. With her father, wearing goggles, seated in front and Jerrie perched on pillows in the rear, they would take the biplane up a thousand feet. Then Harvey Cobb would raise both of his hands to signal that the plane was all hers. Jerrie would take the stick in front of her and move it forward, back, to the right and left. She was astounded at the way the Waco responded to every move she made and she loved how confident flying made her feel.[7]

When the family moved from Denver to Oklahoma City, Jerrie Cobb was a junior at Classen High School. Like much of Oklahoma City, everything about Classen was focused on the space age. The athletic teams were called the Comets; the yearbook was the *Orbit*. Futuristic architect Buckminster Fuller would design a geodesic dome for the Bank of Tomorrow on nearby Route 66. At Classen, Cobb found a kindred spirit who also happened to be a certified flying instructor. Math teacher and Classen football coach J. H. Conger had been taught to fly by famed aviator Wiley Post. Long before Post made the first solo flight around the world, he was an Oklahoma farm boy who gave flying lessons in planes Conger called "old crates I'd be afraid to climb into now."[8] Under Conger's tutelage, Cobb learned to fly the school-owned Aeronca and she soloed on her sixteenth birthday. She worked nights and weekends to earn money for more lessons and earned her private pilot's license on her seventeenth birthday. Her commercial license came on her eighteenth. When she graduated from Classen, all she could think about was buying an airplane and earning her living as a pilot. She even had a plan.[9]

Besides Coach Conger, Cobb found one other reason to go to school: she loved playing on the softball team. When she finished high school, she made a deal with her parents and agreed to delay college for a year if she could play professional women's softball. She began her season with Oklahoma City's Perfecut Manufacturing team before signing with the Oklahoma City Downtown Chevrolet Sooner Queens. In silk skirts and blouses, the women drew loyal fans, sometimes as many as 1,600 spectators for a game.[10] Reporters regularly covered the teams with serious copy about clutch hitting as well as frivolous headlines such as "Softball Back, 'n with Curves."[11] At best, the sports coverage sent a mixed message to women: be aggressive, yet act demure. Cobb took on grimy team jobs such as raking the base paths and cleaning spikes clotted with mud.[12] Yet she also knew she was expected to be acquiescent, soft-spoken, and sexually attractive to men. It would be the same message Cobb followed when she set her three world records in the Aero Commander: fill your plane with oil, get out your tool kit and fix anything that breaks in midair, but make sure your high heels are on and your lipstick is in place when you land. Softball did pay off for Cobb when she was able to buy a used World War II surplus Fairchild PT-23 during an out-of-town softball series in Denver. She convinced the Sooner Queens to front the money for the maroon-and-yellow plane, which she named *Par-a-dice Lost*.[13] Softball was really a means to an end for her, and after countless rides crunched in the back of a station wagon to distant games and a hard-to-forget three-two strikeout with bases loaded in the bottom of the ninth, Cobb ended her professional softball career.[14]

When Cobb returned her focus to school it was Oklahoma College for Women in Chickasha that she chose, not Oklahoma University, which three generations of her family had attended. Cobb's attraction to OCW was not the psychology, sociology, and elementary journalism classes that she attended during her freshman year, but rather the courses in aviation the college offered through the Orville George Flying School. Also there was the school's proximity to the Chickasha Municipal Airport, where she found general maintenance work and some crop-dusting jobs after class.[15] A more advanced pilot than the rest of the students, Cobb bided her time at the college in a single room without a roommate, a dormitory arrangement that she had specifically requested. Despite her well-reasoned choice of OCW, college was often an excruciating time for Cobb. Like many schools for women in the late 1940s, OCW believed preparing young women to be homemakers was its primary mission. "Since the Home is the foundation institution of society," the catalogue read, "the work in social fundamentals is organized around the ideas of homemaking as a vocation." Lectures were offered in personal appearance, voice, diction, cosmetics, dress, conversation, and social adjustment. Speech training, in particular, was emphasized, with each student making a tape recording of her voice for professorial critique and every freshman required to take speech class.[16]

Cobb's speech class ended in disaster. When she came to class unprepared one day, her professor assigned her an impromptu five-minute speech on hats. Cobb began by talking about the only hat she owned—a black one she wore to church. She then described hats aviators used to wear in open cockpits and the caps worn by catchers, umpires, and outfielders—anything that was familiar. One minute passed. Four remained. When the professor asked her to keep going, Cobb refused and walked out of the classroom. Cobb's parents recognized that their daughter was not meant for the Oklahoma College for Women. Cobb's pledge to her parents met, her attempt at college completed, she set her sights on earning a living.[17]

After a few early jobs in Oklahoma City giving flying lessons, Cobb returned to Ponca City, where her father had run an automobile dealership since 1950. "The closest I got to a plane was when one flew by overhead," she remembered.[18] For months, Cobb helped out at Cobb Pontiac and Chevrolet and occasionally flew an oil executive to an out-of-town meeting. Itching for more time in the cockpit and even some friendly competition, she investigated women's air races and decided to enter several derbies. These races, especially the All Women's Transcontinental Air Race (AWTAR), were serious contests that challenged a pilot's skill despite the fact that humorist Will Rogers once dubbed an early race the "Powder Puff Derby" and the name stuck. The races were also long-anticipated social events. Jobs were arranged around air races and family schedules uprooted so that mothers could get away from their children for a while to join other women flyers. Most women flew in teams of pilot and copilot and wore matching outfits. All were members of the Ninety-Nines, the international organization of women pilots founded by Amelia Earhart. (When Earhart sent out a letter in 1929 expressing interest in forming a group for women pilots, ninety-nine women agreed to join. By 1960, there were more than 1,300 individual members and seventy chapters across the country.[19]) At the end of each air race, women who shared a love for flying gathered for festive banquets and casual get-togethers that often lasted half the night. The banquets were memorable and dress was considered extremely important, just as it was in the competitions. Women wore everything from Pendleton jackets with saddle shoes and bobby socks to full-length formal gowns.[20]

Cobb met Dallas pilot and future Mercury 13 member Jerri Sloan at the 1958 race. The two women could not have been more different. Sloan was a great talker, telling funny stories often punctuated by a blush-inducing phrase or two. She had a unique ability to express her uncompromising opinions on all subjects, especially politics, and to put nearly anyone at ease. Even if you disagreed with Jerri Sloan, it was almost impossible not to like her. Although quiet and unassuming, Cobb made an impression on Sloan. "Jerrie was quiet and . . . I am not," Sloan said. "Maybe it was opposites attract." Sloan was right. Cobb liked people who were gregarious and easy to be around. They lifted some of the burden by making conversation for her. "I liked Jerri because of her down-to-earth friendliness," Cobb later said, her "tell-it-like-it-is honesty, Texas-at-its-best humor, and devotion to flying."

Three years after they first met at the Powder Puff Derby, Cobb made sure Dr. Lovelace called the energetic Jerri Sloan and invited her to Albuquerque for astronaut tests.[21]

In Cobb's first air race, she flew solo in the Dallas-to-Topeka Skylady Derby and came in third place. She then took on a cross-country race—again solo—from Santa Ana to New Jersey and took fourth. In an international race from Ontario to New Smyrna Beach, Florida, Cobb came in fourth again, but in this contest she found something more valuable than a trophy: she discovered Florida. The Oklahoma pilot fell in love with the sun and endless beaches. Florida virtually glittered to Cobb and over the next decades would become the staging ground of many key events in her life. It seemed to Cobb as if every time she set foot in Florida, she found opportunity. On that first visit fellow pilots assured her that Miami was America's real gateway to aviation.[22] She started looking for a job immediately.

When Jack Ford stepped up to the Aerodex customer service counter at the Miami International Airport, he did not expect to find his newest employee in the unobtrusive woman helping him with a repair order.[23] Ford's company, Fleetway Inc., an international aircraft ferry service out of California, scouted out war surplus planes, refitted them, and sold them to customers around the world, such as the Peruvian Air Force. He needed pilots who would ferry planes to distant locations, work alone, eat food they sometimes did not recognize, sleep anywhere they could find dry, safe conditions, hop commercial flights back to the States, have a post office box rather than an apartment or a house, have no family ties that required a reliable presence, and be willing to start the flying cycle all over again every ten days. Jerrie Cobb, working behind the counter, fit the bill.

As a tryout for employment, Ford asked Cobb to follow him down to Peru in a T6. When Ford's plane needed new parts in Colombia and was grounded, Cobb continued on alone until the Ecuadoran Army seized her plane and threw her in jail on suspicion of being a spy. Days later, Cobb finally received government clearance and was allowed to take off again.[24] With political tensions simmering between Ecuador and Peru and Cobb flying an airplane viewed as a Peruvian military aircraft, she was fortunate to be released. The anxious incident proved her competence, however, demonstrating that she could navigate in unfamiliar physical and political terrain and remain calm, not easily provoked to rash action or unnecessary moves.

On later flights to South America, Cobb seemed fearless and never complained, often sleeping in her plane with a tarpaulin covering her to keep out the rain.[25] Her self-reliance and steady, even immotive, nature was an asset in tight situations, and she demonstrated extraordinary patience. Within a short period of time, Cobb was flying solo all over the world for Fleetway in an ever-increasing variety of aircraft, including B-17s, C-46s, DC-3s, and amphibious PBY-5As. "I soon became used to being an oddity—the North American lady who flies airplanes—and roughing it was nothing new," she said. "I was conditioned to spending long hours without rest or company, keeping a fast pace, finding my way about in strange places."[26]

Work at Fleetway was a dream job for Jerrie Cobb. For someone whose first trip across the continental United States had been the Santa Ana-to–New Jersey air race, visiting international cities from Europe to India was an eye-popping adventure. She bought llama rugs and collected coins from every new country that she visited. She was equally pleased with her family's pride in their globe-trotting youngest daughter. Cobb discovered a way of life that suited her, one characterized by solitude and motion. She would spend her days by herself, floating in a condition that she once described as "a wonderful state of silent aloneness."[27] Stopping on the ground long enough to take on food and gasoline or transact Fleetway business, she then would sail above the world again. She was perfectly satisfied. Soaring over the tropical rainforests of South America provided Cobb with the same peace she first felt when soloing back in Oklahoma. Looking back, Cobb later recalled that "for a child who distrusts ordinary, everyday speech, for an adolescent who yearned for the freedom of the fields and the winds . . . for a girl who learned to be alone—the sky was the answer."[28]

Work with Fleetway was a rare opportunity for a woman pilot. Cobb was able to rack up hundreds of hours in her logbook and gained unique experience flying a variety of planes in unpredictable weather over challenging and sometimes threatening areas. But Cobb gave up the job in the fall of 1955. She and Jack Ford had begun a romantic relationship that Cobb described as "full of surprise meetings in exotic places."[29] But the romance did not last; Cobb loved flight and solitude more. She looked at marriage between pilots as "no glamour—only frustration, anxiety, and a feeling of deprivation."[30] Four years after she left her job at Fleetway, Cobb learned that Jack Ford had been killed on Wake Island. His plane had exploded just after takeoff. Jack Ford was a pilot so intent on aviation perfection, *Flight* magazine later wrote, that he became "out of place and impractical anywhere but in the air."[31] The passage could have been describing Cobb.

After leaving Fleetway, Cobb retreated to Ponca City once again to consider her next step. A national wire service decided to commission a personality profile story on the young woman who was drawing attention as an international ferry pilot. It assigned Oklahoma newspaperwoman Ivy Coffey to do the story. Coffey made an appointment and went out to Cobb's parents' neat sandstone home on Cleary Avenue. The interview was "the most difficult one I ever had," Coffey wrote. Cobb was monosyllabic; her answers "almost amounted to rebuff." Her manner puzzled Coffey, who tried putting down her pencil and notebook to ease Cobb's anxiety. After the fifth question, Coffey realized that Cobb was not being rude, but was "modest and was reserved to the point of being shy." Finally Cobb's mother pitched in, doing what she could to draw out her daughter. The interview "limped to a finish."[32] Nevertheless, the meeting was hardly a failure. For some of the same reasons Cobb enjoyed sociable Jerri Sloan, she also liked Ivy Coffey. With uncharacteristic forwardness, Cobb later invited Coffey to lunch and thanked her for the story, apologizing for being unable to talk easily. The two women became friends and years later Cobb astonished Coffey by saying she "knew her better than anyone else." Ivy Coffey thought Jerrie Cobb was an enigma.[33]

Cobb was becoming a world-famous aviator. In France, the Fédération Aéronautique Internationale presented Cobb with its gold wings award; she thus became only the fourth American to receive the honor. The Aero Design and Engineering Company in Oklahoma City took note of the way Cobb drew worldwide publicity and attention to their sleek Aero Commander aircraft. Tom Harris, a former American Airlines executive and vice president and general manager at Aero Design, offered Cobb a permanent position with the company as an executive pilot and manager of advertising and sales promotion.[34] Like the rest of the aviation world, Harris was beginning to think that Jerrie Cobb was becoming so well known, she might become the next Jacqueline Cochran. And everyone knew that Jackie Cochran was the most famous woman pilot in the country.

At fifty-three—or so most people thought, since she often professed to be younger—Jacqueline Cochran had achieved every accomplishment a woman pilot could attain: world records, presidential citations, World War II glory, being president of Amelia Earhart's Ninety-Nines, wealth, fame, access to power. No other person received the same treatment that Jackie Cochran did when she walked into a room of women pilots: everyone stood up. Many women snapped to attention out of respect, but most stood out of fear.[35] Jackie Cochran had been relentless in her pursuit of aviation fame, and that often meant bulldozing anyone who stood in her way. As much as she was a standard-bearer for women in aviation, she was not a strong supporter of individual women. She opened doors to female pilots so that she could be the first to walk through them. As long as she was number one, Cochran cared little about who followed; they had just better stay far enough behind. Although Jackie Cochran achieved her greatest fame flying propeller aircraft, she showed no signs of stopping when aviation entered the jet age. Following Alan Shepard's historic launch, a reporter asked Cochran if she had any future plans for spaceflight. "I can't imagine," she said, "that the space age will pass me by."[36]

No one—not her husband or perhaps even Cochran herself—knew the complete factual history of Jacqueline Cochran. Her birth date was never confirmed, although it was estimated that she had been born in 1906. Raised by a foster family, Cochran eventually hired a private detective to find out the identity of her biological parents. The detective prepared a report and handed it to Cochran in a sealed envelope. Cochran passed the envelope on to her husband, Floyd Odlum, who never opened it; Cochran's friend Chuck Yeager reported that the envelope had been burned unopened after her death.[37] What is certain is that she was named Bessie Mae Pittman and grew up in a poor lumber town in northwest Florida. "Sawdust Road," she called it in her 1954 autobiography, *The Stars at Noon*, which many critics saw as a book more intent on fashioning a legend than conveying facts. From the start, the young girl showed exceptional determination to improve her condition and create a dramatic persona, including changing her name to Jacqueline Cochran—a surname she found by reading the Pensacola telephone book. She left Florida with little or no formal education and studied briefly for a career in nursing. She then admitted that she could not stand the sight of blood and found work instead in southern beauty shops.[38] Still working as a hairdresser, she moved to New York, found a job at Saks Fifth Avenue, and cultivated a posh clientele. There she met Odlum, the wealthy head of innumerable corporations, among them RCA, General Dynamics, the Atlas Corporation, and several aircraft-manufacturing firms. In 1935, she developed her own line of cosmetics and founded her own company, becoming the chief executive of Jacqueline Cochran, Inc. A ride on a friend's plane thrilled her, and within a short period of time—her autobiography recounts it was three weeks—she earned a pilot's license and began conducting her cosmetics business by flying around the country. Cochran and Odlum married in 1936, and Floyd became an enthusiastic supporter of her involvement in air racing. "Going for speed records costs a fortune, but [Floyd] happily paid the bills," Chuck Yeager wrote.[39] In 1938, Cochran won the prestigious California-to-Cleveland Bendix race, speeding across the finish line with only minutes of fuel left in her gas tank, a field of male pilots behind her. Later, Cochran's wealth, World War II record, powerful connections in the military, and ferocious arm-twisting allowed her to borrow Air Force planes for record-breaking flights, including crashing through the sound barrier in a jet in 1953 with Yeager chasing. Yeager, more than anyone else, knew the power of Cochran's determination. Jackie Cochran, he said, "would not stand for anyone cutting her props out."[40]

Cochran's access to the military developed from her tireless work organizing the Women's Airforce Service Pilots, the WASPs, during World War II. Concerned that the approaching war would severely deplete the pool of male pilots and seeing a way that she could create a position for herself, Jackie Cochran approached the Army Air Corps with the idea of using civilian women pilots to ferry military aircraft around U.S. air bases. Cochran proposed leading a group of women pilots who also would engage in broader aviation work, including towing aerial gunnery targets, flying engineering test flights, and supervising check rides for military pilots being introduced to newer versions of planes. The chief of the Air Corps, Henry "Hap" Arnold, turned Cochran down, believing there were adequate numbers of male pilots available. If the need for civilian women pilots ever developed, Arnold promised, he would see that Cochran got a position directing the effort. Rebuffed, Cochran left for England, where she studied the way British women pilots assisted the military and set up her own small group of American women aviators who ferried planes for the British Air Transport Auxiliary—the "ATA girls," they were called. When Cochran returned to the States, she discovered that another woman pilot, Nancy Harkness Love, was already organizing an Army Air Corps women's ferrying division called the

WAFS, the Women's Auxiliary Ferrying Squadron. Cochran was furious. She believed General Arnold had betrayed her and Nancy Love had plotted behind her back. Arnold claimed there had been a breakdown in military communications and tried to appease the dueling women by awarding Cochran leadership of all civilian women pilots assisting the newly renamed Army Air Force. In 1943, the entire group of women pilots was renamed the WASPs, with Jackie Cochran serving as their head and responsible for training them to fly "the Army way." Love was handed authority over the smaller women's ferrying division. Cochran had authority over Love, but the two women operated autonomously, mainly by steering clear of each other.

More than twenty-five thousand women pilots applied to become WASPs, and Cochran accepted nearly two thousand of them. Trainloads of women pilots began making their way to Avenger Field in Sweetwater, Texas, where they learned to fly seventy-seven different kinds of aircraft, including planes with dangerous reputations, such as the B-26—the infamous "widow maker"—and the massive four-engine B-24. Over the next year, WASPs logged more than 60 million miles flying military personnel, medical supplies, and cargo all over the United States. In 1944, as many male pilots returned to stateside duty, the Army Air Force deactivated the WASPs and abruptly sent the women home. As much as the women pilots regretted losing their chance to participate in the war effort, they were more incensed by Washington's refusal to recognize them as veterans. The government maintained that the women were civil service employees, not military personnel. Without military status, the WASPs could not receive military benefits, including access to VA hospitals, the GI bill for education, or even bus fare home after deactivation. Discrimination against the WASPs was just as egregious during the war, when they had had to take up collections to send home the bodies of thirty-eight pilots killed in the line of duty. Not only did the United States refuse to pay the expenses of shipping home the women's bodies, but it would not permit American flags to be draped on their coffins.*

---

*In 1977 Congress granted the WASPs veterans status. Two years later, the secretary of defense ruled that their World War II efforts had been active military service. Yet the WASPs' contributions still were not completely recognized. As recently as 2002, the daughter of a recently deceased WASP discovered that her mother was not entitled to full military funeral honors at Arlington National Cemetery. In an eloquent op-ed piece in *The Washington Post*, Julie I. Englund described how Arlington maintained that WASPs did

Some of the WASPs blamed Jackie Cochran for not doing enough to support their military recognition. If the WASPs were recognized as military, some people argued, Cochran would have had to report to the female military head of the Women's Army Corps, the WACs. While Cochran often served as a peer of or even subordinate to men, she never saw herself as any woman's equal and certainly not as her inferior. Other WASPs, such as Margaret Boylan, criticized Cochran for not supporting her pilots once her position as their leader ceased to exist. In looking back, Boylan believed Cochran "could have kept us going." She "just wasn't willing to pay the price it was costing her. And she was the type of person to make a cut-and-dried decision. If she couldn't do it her way, she'd just as soon not do it anymore."[41]

Jacqueline Cochran's reputation for determination, domination, and self-promotion intimidated many people, including politicians, generals, businessmen, and many hotshot pilots. Dr. W. Randolph Lovelace II, however, was not among them. He respected Cochran's fierce will, courage, and skill as a pilot. In fact, Randy Lovelace considered Jackie Cochran one of his closest friends. He and his wife, Mary, named their youngest daughter Jacqueline and asked Cochran to serve as the child's godmother.[42] In Randy Lovelace, Jackie Cochran saw a man whose passion for flying matched her own, a man known for his vision and restless energy. "Don't tell me how difficult it is," a Lovelace colleague remembered him saying. "Just tell me when you get it done."[43] Cochran admired Lovelace's commitment to the future of aviation and expected him to help her be part of it, including spaceflight.

Jackie Cochran first met Randy Lovelace in 1937 at the National Air Races in Cleveland and was intrigued by the young physician's research on high-altitude flying.[44] Cochran was already setting altitude records and searching for a way to push her plane higher without blacking out. In the late 1930s, collaborating with Dr. Walter Boothby and Dr. Arthur H. Bulbulian at the Wright Aeromedical Laboratory, Lovelace developed an oxygen mask for high-altitude flight. The BLB mask, named after the initials of its creators, would save the lives of many World War II airmen who were

not qualify for the same honors as men. Her mother, Irene Englund, was entitled to treatment given a veteran's spouse but did not qualify, they argued, for a military honor detail, a rifle salute, a rendition of "Taps," and presentation of the American flag. Following publication of the op-ed piece and public response in support of full Arlington recognition for WASPs, Arlington reversed the decision.

forced to bail out at extreme altitudes.[45] In 1940, President Franklin Roosevelt, under intense lobbying from Cochran, awarded Lovelace and his colleagues the Collier Trophy, which recognized the year's most significant aviation achievement. But what Lovelace became recognized for—an episode that he reportedly grew exasperated hearing recounted—was his fabled parachute jump.

Randy Lovelace wanted to determine if an emergency oxygen mask could enable a pilot to survive a high-altitude bailout as well as high-altitude flight. He decided the only way to find out was to jump himself. Around noon on June 25, 1943, while on a consulting trip for the Boeing Aircraft medical department, Lovelace jumped out of a B-17 bomb bay eight miles above Ephrata, Washington. It was the highest parachute descent ever attempted in the United States and quite possibly in the world.[46] The leap was Lovelace's first and last parachute jump, and it nearly killed him. A new theory was being advanced that a chute opened at high altitude would jolt a parachutist less than one opened farther down in heavier air. But pathologist Tom Chiffelle, who would later work with Lovelace in Albuquerque, saw horrific injuries suffered by pilots who attempted to bail out of planes at high altitudes. With the advent of jet engines and faster planes, Chiffelle examined pilots assaulted by the blast of high-velocity winds that wrenched the body, fractured limbs, dislocated joints, and deformed the head, trunk, and neck.

Lovelace could recall nothing after the parachute's snap. "I can remember seeing the Fortress just above me," he recounted. "The noise of the engines was very faint. Then the static line yanked out my chute. I felt a tremendous jerk and saw my leather glove and silk glove underneath it both fly off my left hand. It seems rather like a dream now, seeing those gloves shooting through space. All the air in my lungs was knocked out, and I blacked out at the same time." By opening the chute at high altitude, Lovelace was hit with as many as eight Gs and was slammed unconscious. When his left gloves were torn off, his surgeon's hand was exposed to a temperature of forty degrees below zero. All that the crew of the Fortress could see as they circled the chute was a limp figure being swung wildly from side to side. "Maybe he hit the plane as he fell," one airman surmised. "I think he killed himself," another said. But the bailout equipment worked, and slowly oxygen streaming through the mask restored Lovelace's consciousness as he plummeted. Lovelace managed a weak wave to those on the ground and landed with a hard bounce before crumpling into the heap of his feather-filled flight suit. He was alive, but barely able to move. All Lovelace could remember of the landing was trying to move his frozen hand toward the warm sunlight.[47]

Secrecy about the jump was important to Lovelace. He kept the news from his wife as well as from Jackie Cochran and her husband, Floyd Odlum, in whose home he had stayed the night before the jump. "He explained the new apparatus and how it would work in an emergency," Odlum said. "But he kept it secret that he himself was going to try it the next day."[48] As confidential as Lovelace was about the jump on one level, he saw the value of press attention. Reporters from Boeing and *Life* magazine were on hand to witness the event, recording multiangle photographs of Lovelace in his bulky flight gear, sitting in the bomb bay, and sailing halfway through the jump—a lifeless dark dot against the broad Washington sky. Quite after the fact, General Hap Arnold awarded Lovelace the Distinguished Flying Cross for the jump.[49] It would not be the last time Lovelace used the media to scoop a story before notifying his superiors. His being proclaimed a hero or a celebrity in the press made it much more difficult for authorities to be critical about experiments after the fact. Lovelace followed the maverick's dictum: it is easier to ask for forgiveness than to ask for permission. Lovelace's friend and X-15 pilot Scott Crossfield understood and admired his daring. "We broke the law every time we accomplished something," Crossfield said.*[51]

For many of the physicians recruited to his medical foundation in the 1950s, Randy Lovelace was the primary reason for moving to Albuquerque. It was not so much that Lovelace was a superb physician or an aggressive administrator or even a well-connected leader. What most impressed his staff was his ability to be a catalyst: to gather capable people, inspire them, and

*A. Scott Crossfield served with Donald Flickinger and Randy Lovelace on the first government group to study human factors (biological, psychological, and sociological effects of human spaceflight) and training of future astronauts and was a member of the National Advisory Committee for Aeronautics (NACA), the forerunner of NASA. As a pilot, Crossfield made history in 1953 when he broke Mach 2. In 1959, Crossfield flew the X-15, the nation's high-altitude, high-speed, rocket-powered research aircraft, to the fringes of outer space. Though among the top echelon of pilots, Crossfield never wanted to become an astronaut himself. "I would turn off the radio [in the aircraft] if I didn't like the help I was getting from the ground, and the medicine men that were running the program thought that was too independent. They wanted medical subjects, not pilots."[50]

fuel their curiosity with unique—at times, historic—work. Dr. Donald Kilgore remembered coming to Albuquerque for a job interview and being astonished by Lovelace's high-octane enthusiasm. On the day of the interview, Lovelace took Kilgore out to lunch at the Kachina Room at the Albuquerque Airport. There he pointed to a sleek plane sitting on the runway. "You see that Aero Commander?" he asked. "That belongs to J. D. Hertz on our Board of Trustees, and he leaves it here eight months out of the year. See that Navion, that's the clinic's airplane." Lovelace said all the right things to the young doctor who had developed a love for flying while in the Navy. While not exactly manipulating prospective employees, Lovelace knew that they had come to survey his clinic because they were interested in pursuing unusual assignments. Kilgore looked at the available planes waiting on the tarmac and made a quick decision about where he wanted to spend the rest of his life. "I was sunk," he said.[52]

One of the reasons NASA chose Randy Lovelace for the job of medically evaluating the seven Project Mercury astronauts was that his Lovelace Foundation had a reputation for maintaining secrecy. In the early 1950s, with the tension between the United States and the Soviet Union high, Eisenhower gave top priority to developing an airplane that could fly virtually undetected over the Soviet Union. When the plane was ready for its first flight, Lovelace was consulted about medical backup for the test pilot. Then he summoned Kilgore. "Would you like to do an interesting project?" he asked, divulging no details. Lovelace instructed Kilgore to fly to Los Angeles aboard TWA, where he would be identified by a black satchel resting on his lap. A stranger met him in LA and told Kilgore to report to Lockheed's gate at 5:30 the next morning. At daybreak, Kilgore and his black medical satchel boarded a DC-3. He looked around at the faces of those who accompanied him: test pilots, he thought. The plane flew a long time, with Kilgore noting the movement of the sun out the plane's window, calculating the direction in which he was headed. East, he thought. Finally the plane landed in a bleak, remote test facility, miles from nowhere in the Nevada desert—a secret government location known as Area 51. Kilgore got out and was ushered to a hangar situated on the edge of a dry lake bed. "Wanna see the bird?" someone asked. Inside the hangar was the strangest plane Kilgore had ever seen: the U-2. Kilgore flew as medical support on this first test run of the surveillance aircraft, in case the "airplane came to grief," he later said.[53] Like his boss, Kilgore never spoke of the experience until top secret details of the plane's creation finally came to light years later.

Randy Lovelace's inclination for taking bold leaps and keeping secrets led him directly into conversation with Donald Flickinger about testing women pilots for astronaut viability. The two men were curious, from a scientific standpoint, about women's physical and psychological potential for spaceflight. A trip Lovelace and Flickinger made to Russia during the summer of 1959 further convinced them the Soviets were serious about using women astronauts. There was not the same kind of prejudice against women in the Soviet Union, they observed. In fact, Lovelace discovered that 70 percent of the physicians in Russia were women.[54] When the two men happened to meet twenty-eight-year-old Jerrie Cobb later that fall at an Air Force Association meeting in Miami, they immediately recognized that the young woman was just the test subject they had been hoping to find. Cobb was young, accomplished, and motivated, and, like them, she was willing to take chances. She only had to hear the barest outline of their testing plans before eagerly agreeing to report to Wright-Patterson and the Lovelace Foundation for examinations. Flickinger and Lovelace told Cobb that they would review her aviation credentials immediately and then contact her with precise testing dates.[55]

It had been just a little over fifty years since the Wright Brothers' plane had first stalled, then sunk, then soared for a full twelve seconds above Kill Devil Hills, North Carolina. People such as Randy Lovelace and Don Flickinger never quite lost their sense of unabashed amazement at the speed with which American flight had progressed. They also never abandoned their eagerness to be a part of it. The same was true for Jerrie Cobb and Jackie Cochran. Although they did not know it yet, Cobb and Cochran were a little like Orville and Wilbur pitching a coin to see who would push the other one off the hill. Only this time, the shove would not be quite as uplifting.

CHAPTER THREE

Girl Astronaut
Program

AIR FORCE BRIGADIER GENERAL DONALD FLICKINGER HOPED TO move quickly in testing Jerrie Cobb. Since he had been responsible for designing spaceflight simulation evaluations for all seven Project Mercury astronauts at Wright-Patterson Air Force Base in Dayton, Flickinger wanted to test Cobb on the same drills and compare her scores to the ones the men had posted the previous spring. While the first phase of astronaut tests administered by the Lovelace Foundation focused on a candidate's physical aptitude for spaceflight, the second phase, conducted shortly afterward at Wright-Patterson, measured how a potential astronaut might respond to the unique stresses of outer space. Under Flickinger's direction, the staff at the Wright-Patterson Air Development Center's Aeromedical Laboratory provided an extensive battery of psychological exams, including a sensory isolation test calculated to gauge how an astronaut reacted to the simulated silence and stillness of space. Tests at Wright-Patterson also measured the astronaut's ability to function in weightlessness and under the heavy gravitational forces of the centrifuge.*

*Between 1957 and 1958, then Colonel Don Flickinger tapped the Wright-Patterson Aeromedical Laboratory as the center for Air Force efforts to develop a "Man in Space" (MIS) program. At the time, before Eisenhower decided NASA would be a civilian agency, the Air Force was still jockeying with the Navy for control of the nation's space pro-

Don Flickinger had first approached NASA with the idea of testing women for their viability as astronauts in the late 1950s. NASA was not interested. The space agency believed women were physically incapable of handling the demands of space. "You must remember," Flickinger's friend Dr. Stanley Mohler* explained, "the aviation journals in the late 1950s were full of articles claiming that when women menstruate, their brain changes, they become distracted and can't think clearly. They're more likely to crash."† Flickinger did not pay any attention to those myths. Rather than dropping the proposal or continuing to lobby NASA, Flickinger, along with Randy Lovelace, decided to test a woman candidate as part of their own independent experiment, not under the official auspices of NASA. If their results proved that a woman scored well on the same tests that the Project Mercury astronauts underwent, and using the same equipment, Flickinger again would approach NASA with the data.[1] In testing first and presenting data later, Flickinger was hoping for an end run around NASA. Scientific data that refuted prevailing social attitudes had to be acknowledged, Flickinger and Lovelace believed. How could NASA ignore the possibility of female astronauts if data collected by two members of its Special Committee on Life Sciences proved that women were physically capable?

Like Randy Lovelace, Don Flickinger was an iconoclast. "Threshold doctors," Mohler called them—physicians who worked on the edge of scientific

gram. Under his direction, Flickinger hoped the lab would be able to develop medical standards for designing a manned spacecraft capable of low Earth orbit. When President Dwight Eisenhower decided to create NASA as a civilian agency, the Air Force's "Man in Space" initiative did not go forward.

*In the late 1950s, Dr. Stanley Mohler served as a medical officer for the Center for Aging Research at the National Institutes of Health and later as director of the FAA's Civil Aeromedical Research Institute in Oklahoma City. He became acquainted with Flickinger through the Air Force's School of Aviation Medicine in Texas. Mohler now serves as director of aerospace medicine at Wright State University in Dayton, Ohio.

†Women who served in the WASPs were so accustomed to hearing medical doctors voice concern about women flying during their menstrual periods that they developed their own response that allowed them to keep flying any time of the month. When asked by medical professionals, "How often do you get your period?" many of the WASPs responded by saying they were "highly irregular." By pretending to have unpredictable menstrual cycles, the women outsmarted prejudiced physicians and avoided any flying restrictions that the medical myth might prompt. Early in her aviation career, Mercury 13 member Sarah Gorelick Ratley received advice from former WASPs about how to respond to the inevitable "menstruation question" from physicians. She followed the WASPs' suggestion. Ratley now laughs to think about a whole generation of women pilots who were perceived by the medical community as "highly irregular."

advancements and occasionally without direct orders from superiors. Trained as a surgeon, Flickinger demonstrated his daring by parachuting to sites of plane crashes over the Burma-China hump during World War II. Although his commanders would have considered the jumps too dangerous, Flickinger was not the type to wait for "some slick pants, pressed-shirt guy" sitting in a quiet office to approve his every move, Mohler said.[2] After the war, Flickinger concentrated on high-altitude ejection problems and in 1959 served as assistant for bioastronautics at the headquarters of the Air Force Air Research and Development Command (ARDC) at Andrews Air Force Base outside Washington, D.C. During the astronaut press conference earlier that spring, Flickinger had commended the quality of the Project Mercury astronauts. The Russians might have beat us in propulsion, he stated, but "the quality of our human component will be far superior to theirs."[3]

Don Flickinger was concerned that the frenetic pace of the space race prevented leaders at NASA and in the federal government from considering the larger implications of their actions. Decisions were made quickly, often without a full appreciation of their consequences. "I think the pace is horribly fast," Flickinger said. "By and large as a nation, we don't take time to simply sit and contemplate. There are too many demands, there are too many crises going on all the time, and I don't think this is good."[4] Even deciding who should be tested for astronaut potential was a determination made out of expediency, a NASA spokesperson admitted. At the spring press conference introducing the Project Mercury astronauts, a spokesman acknowledged that narrowing the astronaut selection process to military test pilots had been a quick decision made for the sake of efficiency. "It was a purely arbitrary decision," he told the press corps, "because we knew that the records on these people were available. We could run them through the machines and very quickly make first-cut selections."[5]

Preparing for his experiment at Wright-Patterson, Flickinger asked Jerrie Cobb if she could suggest the names of other women pilots who might be qualified in terms of their medical records, age, height, weight, and flight experience. Flickinger hoped to test a group of some ten women so that Cobb's scores would not be viewed as exceptional and therefore not representative of women as a whole. Cobb immediately supplied him with seven names.[6] Flickinger checked their status through the Civil Aeronautics Authority (the forerunner of the Federal Aviation Administration) and proposed eight additional women who might serve as substitutes if for some reason Cobb and her cohorts did not meet his medical and flight experience requirements.* Flickinger especially was interested in candidates who were under thirty-five years old. Had Flickinger applied the same age qualification to the Project Mercury astronauts, three men would have been eliminated: Alan Shepard and Deke Slayton were already thirty-five, and John Glenn was thirty-seven years old.

Jerri Sloan, who was on the list of names Cobb had proposed, said that Cobb had a habit of volunteering her first and telling her later.[7] Sloan did not have to worry about receiving a surprise call, however. Neither she nor any of the other women pilots whose names were suggested would ever be invited for Wright-Patterson tests. No one but Randy Lovelace and Don Flickinger knew that another woman pilot was already undergoing astronaut exams at the Wright-Patterson Aeromedical Laboratory. Several weeks later, public disclosure of Ruth Nichols's private tests sparked the public's interest but prompted consternation from the Air Force. The attention also triggered a decisive Wright-Patterson backlash aimed directly at Don Flickinger.

At fifty-eight, Nichols was a generation older than the Project Mercury astronauts and well beyond the age of the other women pilots suggested for Flickinger's testing. Practically no one would ever have considered Nichols for astronaut viability, but few dismissed her determination. Nichols was Amelia Earhart's principal rival and would have crossed the Atlantic one year before Earhart had she not crashed in Newfoundland. The Newfoundland crackup was one of many for Nichols, but accidents and other setbacks rarely stopped her from pushing ahead. Nor was she willing to admit that age might slow her down. Like Jackie Cochran and Jerrie Cobb, Ruth Nichols realized that space was the next horizon for top-notch pilots. She wanted a chance to experience some of the astronaut tests herself and worked with Flickinger to set up a run on Project Mercury experiments at Wright-Patterson Air Force Base.[8]

The practice of allowing civilians, such as Ruth Nichols, to experience

---

*The seven women Cobb proposed to Flickinger were Frances Bera, Barbara Erickson, Marilyn Link, Marian Petty, Betty Skelton, Jerri Sloan, and Jane White. Flickinger's list included Marian Burke, Margo Callaway, Sandra Callaway, Jerelyn Cassell, Evelyn Kelly, Juanita Newell, Aileen Saunders, and Jimmye Lou Shelton.

military life—riding in jets or using space simulation equipment at Wright-Patterson—was not uncommon, especially during peacetime. From 1959 to 1962, the United States was between the Korean War and Vietnam, leaving a surplus of scientists and equipment at places such as the Wright Air Development Center. Stanley Mohler, who worked with both NASA and the military, knew that women could occasionally gain some additional flight experience or testing through the military if officers had "the time, the equipment, the energy and the desire" to provide a onetime opportunity.[9] In fact, at least one other person had already tried out some of the tests given to Project Mercury astronauts. A *Life* magazine science editor participated in some of the same experiments for a feature that appeared just a few months before Nichols took her turn.[10] The Air Force probably viewed Nichols's testing as it had the *Life* editor's—as a useful public relations exercise. In addition, Ruth Nichols's brother, an Air Force colonel, could have opened some doors for her. Nichols's own reputation as a daring pilot with a storied past certainly made the Air Force treat her with respect.

Ruth Nichols started flying as a teenager simply because she knew that if she did not face her fear of being in an airplane, she would be "licked all the rest of [her] life."[11] Before long she became the first woman ever licensed to fly a seaplane. According to her brother, Nichols would arrive for college at Wellesley in a flying boat that she would "land on the lake on campus, taxi to the dock," then, after joining the college crew team for practice, she would jump back in her seaplane and return home to New York.[12] Nichols went on to pilot over 140 different aircraft, survive six major crashes, and eventually set women's records for speed, altitude, and nonstop heavy airplane distance flying. Like Jackie Cochran, whose wealth provided her with rare opportunities, Nichols's privileged New York and Palm Beach life and personal contacts also opened doors that were closed to most other women pilots who wanted access to jets and, eventually, space capsules. Among her friends were the Firestones, the Goodrich family, and Charles and Anne Morrow Lindbergh. Charles Lindbergh told Nichols he "didn't think it was such a hot idea" for a woman to attempt flying the Atlantic solo but nevertheless helped Nichols plan her flight from Newfoundland to Ireland as they sat before the fire at the Lindbergh home.[13] Amelia Earhart was also a friend. The two competed good-naturedly for records and exchanged remedies for dealing with queasy stomachs when occasional severe air turbulence made them nauseated. For

Earhart the cure was tomato juice. Nichols preferred caviar sandwiches. "They do have a great deal of nutritive value," she said.[14]

The night Nichols departed from Floyd Bennett Field on the first leg of her proposed solo flight across the Atlantic, her heavily loaded Vega—named *Akita*, the Sioux Indian word for "explore"—was barely able to get off the ground.[15] In Newfoundland, blinded by a setting sun, she skidded off the short runway into some trees, crawling out of the wreckage with five broken vertebrae. Only three months later, wearing a steel corset, she went after the women's distance record in the rebuilt plane. She convinced her orthopedic surgeon to invent a contraption to assist her injured back, and mechanics rigged two straps to the top of the cockpit to hoist her up. By slipping her arms through the straps, Nichols could ease the weight off her spine and make the long flight from Oakland to Louisville more bearable. Nichols broke the record.[16] The next day, while she taxied the Vega, the plane caught fire. Nichols escaped injury, but extensive mechanical repairs needed to be made on the aircraft. Despite financial support from the Life Saver Company, the repairs were time-consuming, and while Nichols waited, Earhart ventured the second Atlantic crossing by a woman.[17] In May 1932 Earhart made the flight that put her name into the history books.[18]

Nichols's long and celebrated career as a pilot also was marked by anger on behalf of women pilots denied opportunities that were handed to their male counterparts. Nichols protested, for example, when pioneering aviator Helen Ritchie was prohibited from continuing to work as the first woman to fly for a passenger airline. "She was forced out of that type of flying by the men," Nichols said. "The pilots' union said that all the pilots would go on strike if the airlines continued to hire a woman as a pilot. I thought it was outrageous . . . and I sent off a wire to both Helen Ritchie and the airline, hoping that they'd be able to keep her on, because it did seem to me that it was high discrimination."[19] A descendent of Quakers who had fought as abolitionists and suffragists, Nichols chafed when men told her that they were protecting women from dangerous situations. Discrimination has nothing to do with chivalry, she said.[20]

Nichols made the most of both her talents and her connections. After the war, Nichols helped organize the civilian group Relief Wings, a forerunner of the Civil Air Patrol, which aided the Air Force during times of natural disasters or other emergencies. Then she turned her sights on jets, an aspiration only well-connected women pilots such as Jackie Cochran could ever hope to attain. With her brother, Air Force Colonel Erickson "Nick" Nichols, watching her, she flew in a jet along with military personnel at Hamilton Air Force Base in California.[21] Later, with Suffolk County Air Force Base commander Fred Hook, Jr., and with Pentagon approval, fifty-six-year-old Nichols flew more than 1,000 miles per hour at 51,000 feet in a Delta Dagger supersonic fighter interceptor. Hook said he was at the controls "only when we were on instruments in the overcast and on landings and takeoffs. After that she took a whirl and handled the plane like a veteran." The flight was like "riding in a limousine," Nichols said, and newspapers declared that she set a new record, besting the previous women's record for speed and altitude set by Jackie Cochran.[22] But the flight was more than a chance to set new records; it enlarged her perspective. "It was a rare experience," she explained. "When you're up that high, and you see how small everyone is below, you have no doubt that everyone has the same amount of rights in the world."[23]

Ruth Nichols began what she called "sampl[ing] some of the astronaut tests" in mid-August 1959, just weeks before Jerrie Cobb's chance meeting with Randy Lovelace and Don Flickinger in Miami. Nichols's tests spanned three days at Wright-Patterson and included weightlessness, isolation chamber, and centrifuge tests.[24] When a K-135 plane used for weightlessness testing was not available, Nichols climbed onto the rudimentary simulator—a mechanism she described as a "platform suspended off the floor by jets of steam underneath." The surface was extremely slippery and made walking feel as though you were sliding on ball bearings. "Any slight motion that you made would skid you off this terrific long room on your ear." Technicians handed Nichols a fifty-pound gyroscope that she was supposed to use as a steering mechanism, controlling the tilt of the platform. Nichols observed that the technicians were more concerned about the safety of the gyroscope than they were about her. "I could hear each one say to the other, 'Now you be sure and grab the gyroscope when she goes, because she's going to go in one direction, and the gyroscope's going to go in the other.' " Nichols found it easy to maneuver the gyroscope, and after years of feeling temporary weightlessness in seaplanes, or "flying boats," as she called them, the simulator was not challenging, in her estimation, and she judged the experience "quite a fiasco."[25]

Nichols reported that the isolation chamber at Wright-Patterson was the toughest test. "I don't think the average person realizes what blackness is, until you're in a place of that sort," she said.[26] The chamber was a dark-ened, soundless, ten-by-twelve-foot room. A number of male pilots who spent only a few hours in the chamber responded violently, Nichols reported. Doctors told Nichols that "those who have any particular psychotic reactions are apt to go completely to pieces under such conditions, because you are so disorientated and unrelated to anything."[27] While no official record indicates how long Nichols was in the isolation chamber, she described her duration as "a short time." The experience reminded her of a commercial plane crash she had survived a decade earlier when adjusting to her environment was a matter of life and death. On a flight from Rome, the pilot overshot Shannon, Ireland, ran out of gas, and ditched in the Atlantic.[28] When the plane hit the water, Nichols recalled the horrifying scene: "suddenly primeval chaos—mountainous waves—blackness."[29] She floated in the Irish Sea on an upturned raft until a trawler spotted the crash debris the next morning and rescued nearly all the passengers and crew.[30] As she waited to be rescued, Nichols tried to remain calm and become aware of all that was around her—water, wreckage, heartbeat, temperature, wind. She held tight to the personal articles she had stuffed in her shirt as the plane went down: four codeine pills, a traveler's check, a pack of Life Savers, and her pilot's license.[31] "My experience probably helped me adjust to the isolation chamber," she said. In the Wright-Patterson chamber, she took in her surroundings just as she had down in the Irish Sea. She clicked her nails against the side of the room and discovered an echo; she located several tables and chairs.[32] Concentration, she admitted, was difficult, as it also was for the Project Mercury astronauts who, like John Glenn, spent three hours in the Wright-Patterson chamber.

Nichols also completed several other tests, including a centrifuge whirl that she said reminded her of a roller coaster that grew more tolerable the more practice she had and space-walk experiments testing handholds on the outside of a capsule mockup.[33] At the conclusion of her three days, the fifty-eight-year-old pilot urged Air Force personnel at Wright-Patterson to use women in spaceflight. Nichols remembered that they responded to her sug-gestion with horror and declared that "under no circumstances" would women be allowed to be astronauts.[34] Apparently Nichols had overstepped the line, voicing her opinion instead of politely thanking the Air Force per-sonnel for their courtesy on her way out the door. Nichols was incensed by the reaction to her idea. The rationale for excluding women from the astronaut corps, Wright-Patterson scientists told her, was that they knew very little about how a woman's body functioned. To Nichols, eminent physicians admitting medical ignorance about half the human race "was an extraordinary statement." They explained that since males had been the subjects of aviation and space medicine for the last twenty years, it would require another two decades at least to reach the same understanding about females. Spending so much time studying women also would divert them from their primary work on men.[35] Nichols suggested an immediate and intensive program to put women through the same experiments the male astronauts had experienced. "From every viewpoint," she said, a woman "could hold her own in a space situation and be of tremendous service."[36]

Colonel John Stapp, chief of the Wright-Patterson Aeromedical Laboratory, had long made up his mind about women as astronauts. Females were considerably less equipped to withstand the emotional stresses that accompany spaceflight, he argued. "Economically, the cost of putting a woman in space is prohibitive," he said, "strictly a luxury item we can ill afford." Although the physicians Nichols spoke to admitted their ignorance of the workings of the female body, Stapp offered a quantifiable assessment about women's physical abilities, albeit without citing any specific test results. Women, he said, were physiologically 85 percent as efficient as men of the same weight, size, and age. He was not sure if women were able to maintain effort and motivation in extremely stressful situations and he doubted that women would be able to be objective and offer sound judgment when they were tired or nervous. Women, he said, needed to be protected against exposure to dangerous work. "To expose women needlessly," he said, "to the known as well as the incalculable dangers of pioneer spaceflight would be like employing women as riveters, truck drivers, steel workers, or coal miners."[37]

Ruth Nichols kept talking. In November 1959, the American Rocket Society gathered in Dayton and the subject of women astronauts was on the table. Nichols spoke out about her recent astronaut testing and reiterated that she thought women were better equipped to handle the rigors of spaceflight than men. Like most other men and women of her time, Nichols believed that women were by nature quiet, dutiful, and models of forbearance. These female characteristics, she argued, would make women excellent astronauts. They would not take rash actions, would respond well to directions from the ground, and would be able to patiently withstand discomfort, isolation, and inactivity. "Most women are passive, submissive, patient by birthright," she said. A constitution not prone to activity would hold up well under long space voyages, she suggested, adding sarcastically, "I think it is significant that every live animal sent up so far has been female."[38]

Perhaps it was Nichols's acerbic tone or her lack of equivocation that Air Force brass found infuriating. Certainly they believed that there was far too much attention focused on Nichols and Brigadier General Flickinger's ad hoc women's astronaut testing at Wright-Patterson. Regardless of the trigger, military officers had heard as much as they wanted to about women astronauts. By December, Flickinger's "girl astronaut program" was scrapped before any additional women could follow Ruth Nichols to Dayton for testing. According to Flickinger, there was too much opposition from Wright-Patterson military men and too great a fear on their part that publicity about women astronauts would hurt the Air Force.[39] Writing to Jerrie Cobb, Flickinger observed that the unfortunate media attention on Ruth Nichols had done much to "turn the tide" against Air Force medical sponsorship of the program.[40] Flickinger was disheartened and told Cobb he was even sorrier than she was that the program could not continue. The barriers to move forward, he confessed, were simply too great to be overcome. He notified officers at the newly renamed Wright Aerospace Medical Division and the Air Force's School of Aviation Medicine that the project was terminated.[41]

Later, in a letter to Randy Lovelace, Flickinger was even more candid. Military men and civilian doctors at Wright-Patterson and the Air Force School of Aviation Medicine told Flickinger that "there was too little to learn of value to Air Force Medical interests" to justify the experiment. In particular, Flickinger cited five men who did not believe the "girl astronaut" program should go forward.[42] All were Air Force colonels or generals who held chief positions in aeromedical research or who worked with NASA.* Dr. Stanley Mohler remembers the men who objected as being strong-minded and of a conservative stripe. "These guys were bulldogs," he said.[43]

---

*According to Flickinger, the men who rejected his proposed "girl astronaut" program included two former heads of the Wright Aeromedical Laboratory: Colonel Jack Bollerud and Colonel Otis Benson, who later headed the School of Aviation Medicine. Others included Brigadier General Charles H. Roadman, who started at Wright-Patterson before becoming director of aerospace medicine for NASA's Office of Manned Space Flight,

While admitting ignorance about the capacity of the female body for spaceflight, the Air Force was not curious to learn anything new. Rejecting any experiment that offered insight into women's physiology was not a matter of disinterest alone. Discovering that women were stronger and more physically capable than assumed might also challenge the military's assertion of male strength, bravery, and superiority. As long as military men were viewed as possessing unique physical qualities that permitted them to accomplish dangerous missions, they held a monopoly as jet test pilots and as astronauts on spaceflights. Wright-Patterson officers simply preferred not to know what women pilots could do rather than face the possibility of "adverse publicity" that might accompany new scientific understanding about women.[44] They certainly did not want anyone to know that a fifty-eight-year-old woman could hold her own in space tests designed for virile young soldiers.

There was also a problem with the partial pressure suits worn by pilots and astronauts to survive high-altitude flight. As anyone who has flown in a plane or climbed a mountain knows, human bodies respond differently at high altitudes. Ears pop; some people experience shortness of breath; a few get dizzy as altitude increases. For pilots like Jerrie Cobb who achieved altitude records in unpressurized aircraft, supplemental oxygen was essential. At altitudes above 50,000 feet, human beings need pressurized suits because blood quite literally begins to boil. Gases that are dissolved in the body rapidly come out of solution and bubble up, just as a canned carbonated soft drink fizzes when it is opened. At extremely high altitudes the atmospheric pressure is so low that, according to NASA, "water vapor in the body appears to boil, causing the skin to inflate like a balloon."[45]

But the Air Force designed partial pressure suits (PPSs) for the male body only. Who knew how much alteration would be necessary to adjust the standard suit to the contours of a woman's body? Although the Soviets had managed to reconfigure PPSs to equip dogs for space launches, the American space program found outfitting women with the necessary suits a complicated and expensive task.[46] X-15 pilot Scott Crossfield said that making pres-

Lieutenant Colonel William Turner, who served as chair of the candidate evaluation committee at the Wright Aeromedical Laboratory, and Larry Lamb, who was a civilian cardiologist at the School of Aviation Medicine and worked as a consultant to NASA's Space Task Force.

sure suits did not always require bureaucracy; sheer ingenuity could be used. Parts of the first pressure suit were made in his own garage on his wife's sewing machine, he remembered.[47] David Clark, who achieved prominence for his manufacture of pressure suits, began his career as a lingerie maker in Worcester, Massachusetts. He started making pressure suits in the 1930s, using women's brassieres as a model. "Brassieres were very well engineered," Crossfield said.[48] When it came to the testing of women at Wright, however, no amount of ingenuity would work. In Flickinger's words, Wright-Patterson could not in the end "justify the expense of altering the PPSs to fit the girls."[49] In a way, the Air Force response sounded like a "wife joke" offered up by some tired comedian on *The Ed Sullivan Show*. Why couldn't women be astronauts? Because they had nothing to wear.

Frustrated and disappointed, Flickinger realized there was no chance that he could persuade the Air Force to go forward with the Wright-Patterson portion of the program he and Lovelace had cooked up. His hand had been firmly slapped by the Air Force, and he could not, at this point, reverse what appeared to be an order. Flickinger offered the entire "girl astronaut" program to Lovelace and even volunteered to speak again with officers at Wright-Patterson if Lovelace wanted to try his hand at persuading officers to oversee stress testing for women at the Dayton labs. "I continue to have a keen personal interest in [testing women pilots] and believe it eventually should be done on as scientifically sound a basis as possible. I feel (by instinct perhaps) that if carefully done with a large enough series, there would be some interesting differences between male and female responses."[50] Flickinger passed on the lists of names that he and Cobb had assembled. In closing his letter to Randy Lovelace, Flickinger wrote that he would happily leave Andrews Air Force Base and the Air Research and Development Command in order to come out to the Lovelace Foundation and do the expert psychological evaluation of the women.[51] That arrangement would never happen, however. The Air Force had slammed shut the door that Ruth Nichols had cracked open. But another door opened in Albuquerque later that winter.

The Lovelace Foundation for Medical Education and Research turned out to be an excellent alternative. It was a civilian facility that did not have to report to a military hierarchy or answer questions about publicly appropriated funds. But equally important, the foundation's energy, vision, influential connections, and dedication all emanated from Randy Lovelace, and as long as he was at the helm, his interests were reflected in the work being done. One of the fundamental reasons that NASA relied on Randy Lovelace as chairman of its Special Committee on Life Sciences was that the space agency was in its infancy in the late 1950s and did not have a comprehensive medical unit as part of its own structure. In 1958, when Eisenhower transformed the National Advisory Committee for Aeronautics (NACA) into NASA, T. Keith Glennan was charged with bringing organization and coordination to the new agency. In his comprehensive history of the space age, *This New Ocean,* William E. Burrows wrote that Glennan was a kind of intent orchestra conductor working to blend individual musicians into an ensemble. Glennan "was under enormous pressure," Burrows observed, "to assemble the musicians and then get them to play together under his baton, while the audience—the President of the United States, Congress, the news media, and the nation—watched anxiously because most of them thought the Russians were ahead."[52] During those early years, NASA was still operating out of the Dolley Madison House on H Street in Washington, D.C., with only 177 employees at the headquarters.[53] The Manned Spacecraft Center in Houston, future home of astronaut testing and training, was two years away from creation. During this early period, when NASA "was not much more than an organizational chart," according to Dr. Donald Kilgore, a colleague of Lovelace's, the Lovelace Foundation was "the de facto medical department" for the space agency.[54] NASA hired outside contractors such as the Lovelace Foundation to complete the work it was not equipped or even designed to handle during the early years. Burrows reported that by 1961 contractors were employing 58,000 people for NASA work. The next year the number of contracted workers rose to 116,000. By 1965, it was 377,000 employees.[55] When NASA needed work to be done that required medical expertise, secrecy, and speed, it called Randy Lovelace.

Accepting full responsibility from Flickinger for the women astronaut testing experiment, Randy Lovelace contacted Jerrie Cobb to find out if she was still interested in being the initial test subject. She remained eager. It did not matter to her if she participated in tests in New Mexico or Ohio. She was willing to travel anywhere. Exams would now take place at the Lovelace Foundation in Albuquerque, Lovelace explained. She would spend a week at the foundation, staying in a motel across the street and undergoing the same series of physical tests that the Project Mercury astronauts had taken the year before. Now that Wright-Patterson was off-limits, Lovelace did not know where Cobb might take the next phase of exams—psychological and spaceflight simulation tests. But such obstacles did not faze Randy Lovelace. He always believed that the details would work themselves out. Looking at the crowded schedule for foundation research projects over the next few months, Lovelace spotted the week of Valentine's Day as a time his staff could devote their attention to Jerrie Cobb. The date worked for Cobb as well, and Lovelace asked that she keep her upcoming travel plans secret. Now all Randy Lovelace and Jerrie Cobb had to do was wait and hope that nothing else would make their testing plans backfire.

CHAPTER FOUR

The View from Albuquerque

IT WASN'T HARD FOR JERRIE COBB TO KEEP BUSY WHILE WAITING for Valentine's Day 1960 to arrive. She did, after all, still have her Oklahoma City job at Aero Design and Engineering, and it more than filled her days with demonstration flying, marketing hops, and aviation meetings. During one typical week, for example, Jerrie flew customers around the Southwest demonstrating the Aero Commander aircraft. Then she made a quick hop back to Oklahoma, stopping just long enough to swap suitcases, pack a new change of clothes, and take off again for New York City. At the Waldorf-Astoria hotel, Cobb hobnobbed with aviation executives at the Wings Club annual banquet.[1] The quiet charm she was able to muster for these social occasions surprised some of her friends. Reporter Ivy Coffey thought that Cobb instinctively understood far more about public relations than she admitted.[2] Somehow she had found a way to convert pilot's hangar talk into Manhattan cocktail party chitchat.

Occasionally the public relations and marketing trips that Cobb handled for Aero Design brought her some personal benefit as well. After the New York event, she flew to Panama City, Florida, for the seventh World-Wide Weapons Meet at Tyndall Air Force Base. The competition, commonly called the "William Tell Project," derived its nickname from demonstration missiles that were fired at jet target "apples." Cobb observed military teams vying for championships in air-to-air missile flying and also studied radar control systems that assisted aircraft flying in bad weather. The Air Force at Tyndall saluted Cobb—"Aviation's Woman of the Year," it called her—with something far more important to her than plaques or accolades. It let her fly a jet.[3]

"Jerrie Cobb Flies Jet Fighter" read the headline of Ivy Coffey's story in *The Daily Oklahoman.* Cobb's friend went on to detail how Cobb had been at the controls of the Delta Dagger TF-102A for about forty minutes of an hour-long flight. Along with an Air Force pilot, Cobb flew at 46,000 feet and broke the sound barrier, streaking over the Gulf of Mexico at night at Mach 1.3. Cobb reported that she found the plane easy to control. "There was no special feeling except at times there was a sense of weightlessness but of course, that's possible in other planes."[4] The only new experience was the push of the afterburner, "like a big kick," she said. Writing privately to her women's air-race pal Jerri Sloan, Cobb was less guarded and hinted at the frustration she felt in not having routine access to jets. "I think we're in the wrong type planes," she grumbled, "and will be glad when the jets get into the private and business aircraft field."[5] Cobb did not want to be forced to depend on courtesy, connections, or public relations to get jet time.

Almost as important as the ride itself were the press photographs of Jerrie Cobb next to the cockpit of the jet. Photographs were valuable to her, especially in these settings. With helmet and flight suit, Cobb looked the perfect pilot, standing confidently with one foot up on the ladder leading to the cockpit, and the photos would be recycled again and again in stories about Cobb and were used on the covers of the two books that Cobb later would write about her career. There was only one thing wrong with the image Cobb wanted to project to the public: the ID tag around her neck. "Official Visitor," it read. However permanently Cobb wanted to fix her image as a jet pilot, the military saw it another way. Jerrie Cobb was always a visitor with interloper status.

Unlike Jerrie Cobb, Jackie Cochran enjoyed unprecedented military cooperation when it came to flying the nation's fastest and most innovative aircraft. Yet some pilots, Chuck Yeager among them, refused to see that Cochran's access to jets had as much to do with her influence as her ability. Yeager believed there were no barriers preventing accomplished women pilots from flying jets in 1959 and used Jackie Cochran as an example to prove his point. Cochran flew jets, Yeager reasoned, because she was aggressive and grabbed the opportunity; other women pilots simply did not have Cochran's talent. In describing how Cochran had gained access to jets, Yeager's logic revealed contradictions. "Jackie bought a P-51, and she raced it because she had the capability. Then, when the F-86 came along, when her husband owned part of Canadair, she talked Canadair into hiring her as a civilian test pilot."[6] Yeager himself provided Cochran with opportunities no other woman pilot enjoyed. He helped train Cochran for her sound-barrier-breaking flight in May 1953 and flew an Air Force F-86 to chase her. The Air Force gave Cochran permission to use its base at Edwards as well as its equipment and the fifteen personnel required to prepare the airplane.[7] General Jimmy Doolittle, however, had some anxiety about Cochran using Air Force facilities. He was particularly worried about the possibility of Cochran crashing and the image it would project of a woman pilot injured or killed on Air Force grounds.[8] The public was not used to seeing women put in harm's way, especially with the government's help. He made Yeager assure him that Cochran was ready for the flight. To Chuck Yeager, Jackie Cochran's record-breaking flights proved that women were not barred from flying jets. "There were at least 1,200 other women who may have been just as good as she was as a pilot," he said, "with just as much experience. But when World War II ended, those other women had given up flying. They went back to running their businesses or to being housewives and mothers. Jackie would never give up. She kept right on flying fast planes."[9]

As Jerrie Cobb neared her testing date with the Lovelace Foundation, she wanted to eliminate all distractions so that she could prepare physically and mentally for the challenge ahead. The long commute from her parents' Ponca City home to her Aero Design office in Oklahoma City, for example, was tiresome. Ivy Coffey lived in an old brick and stucco house that she rented on Classen Drive in Oklahoma City and had plenty of room. Since she was away so much covering stories across the state, Coffey offered Cobb space in her home.[10] Cobb moved in, and soon the ten-room home became a training camp for her upcoming testing at Lovelace. Just as he had done with the Project Mercury men, Randy Lovelace asked Cobb not to tell anyone about the trip to Albuquerque. He preferred to release news of the testing once the data had been analyzed. Though Cobb kept her secret, coworkers at Aero Design were mystified when she came into work each morning for six weeks with wet hair and an exhausted disposition. Cobb had devised her own exercise regimen that included running twenty barefoot laps (whenever possible, she preferred to run shoeless) at 5 A.M. around a large vacant lot across from Coffey's house. After a quick shower with no time to dry her hair, she was off to work. In the evening she would repeat the barefoot circles—gradually increasing her distance to five miles—and started riding twenty miles on a stationary bike set up in the den. When time permitted, she sometimes added a round of golf or tennis and swimming to her routine. She also considered her diet and loaded up on foods that, in her judgment, seemed best for sustaining her through the arduous tests ahead. Cobb downed pork chops and steaks, cheese and milk, and continued her habit of hamburgers for breakfast.[11] At night she would stumble with fatigue into her room filled with souvenirs from her days as a ferry pilot—"Do Not Disturb" and "Occupied" signs from South American jetliners and European hotels, an ebony crucifix from South Africa.[12] Finally, collapsing into bed, she would sleep for nine hours, occasionally even more, and begin training again the next morning.

Cobb had an idea of what she was getting into. Randy Lovelace had worked with the editors of *Life* magazine to write a two-page story describing the physical tests he had devised for the Project Mercury astronauts. The *Life* article appeared after the seven astronauts had been publicly introduced and provided Cobb with a sense of what would be required in Albuquerque. Almost as significantly, Lovelace's prose style and the testing details that he focused on provided an insight into his personality. It was easy to see that Lovelace was thrilled to be part of the great space venture and proud that the Lovelace Foundation had been selected for testing the astronaut candidates: "Selected in part," he wrote, "because its geographical isolation simplified NASA's security problems."[13] Lovelace made no attempt to minimize the comprehensive nature of the exams, calling them "one of the toughest medical examinations in history." He offered examples of seventeen separate procedures to examine the eyes and described the uncomfortable motion sickness evaluation and an "unpopular" test that involved planting a tiny electrode into a pilot's hand muscle to measure its electrical response to nerve stimulation.[14] But Lovelace also showed a sense of humor and could not resist noting the incongruity of macho test pilots reduced to an image of basic bodily functions. "Our staff got used to the sight of men walking down the corridors carrying a briefcase full of paperwork in one hand," he wrote, "and a gallon jug containing urine in the other."[15] Evidently Randy Lovelace valued an easygoing, optimistic, uncomplaining personality—both in himself and in others.

Despite her vigorous preparations, one of Lovelace's tests filled Cobb with dread. "What the rider in the first human space capsule has to say will be as important as anything his electronic gadgets report," *Life* reported.[16] The magazine then printed word for word a paragraph designed by Western Michigan University professor Dr. Charles Van Riper that claimed to contain all the speech sounds in the English language. Nearly everyone who saw the paragraph in *Life* read it out loud to measure the effectiveness of his own enunciation. The paragraph was this:

> You wished to know all about my grandfather. Well, he is nearly 93 years old; he dresses himself in an ancient black frock coat, usually minus several buttons; yet he still thinks as swiftly as ever. A long, flowing beard clings to his chin, giving those who observe him a pronounced feeling of the utmost respect. When he speaks, his voice is just a bit cracked and quivers a trifle. Twice each day he plays skillfully and with zest upon our small organ. Except in the winter when the ooze or snow or ice prevents, he slowly takes a short walk in the open air each day. We have often urged him to walk more and smoke less, but he always answers, "Banana Oil!" Grandfather likes to be modern in his language.[17]

With memories not far behind of her childhood speech impediment and the ill-fated speech class at the Oklahoma College for Women, Cobb must have thought no expertise in the cockpit could help her with the exercise.

Cobb could not have known that Randy Lovelace, besides being voted by his high school class "least likely to succeed," was a miserable speaker as a young man. During his senior year, Lovelace's public speaking teacher rated him "poor."[18]

A little less than two weeks before Cobb would depart for her secret trip to Albuquerque, another woman pilot had everybody talking about female astronauts. Betty Skelton, whom Cobb had suggested to Brigadier General Flickinger as a possible candidate for his "girl astronaut" program, was featured in *Look* magazine after undergoing a preview of "a flight into space from a woman's point of view."[19] Skelton's invitation to take part in the feature came from the magazine's aviation editor and was not instigated by NASA or NASA contractors. The feature did, however, have the cooperation of NASA, the U.S. Air Force and Navy, the McDonnell Aircraft Company, and the Project Mercury astronauts themselves. Some detractors called the article a publicity stunt. Certainly none of the cooperating agencies viewed their participation as anything other than public relations. NASA did not see Betty Skelton's *Look* magazine exercises as a prelude to an official women's astronaut program. Skelton herself was realistic about the intent of the article. As a pilot, she had flown in air shows with the Navy's Blue Angels and was keenly aware that she was flying a propeller plane while the Navy men were in jets. She knew there were plenty of men—she called them "skeptics"— who regarded women as less than qualified for aviation challenges such as flying jets and spaceflight.[20] While understanding that she was not being considered an astronaut candidate, Skelton wanted to succeed and prove that women were up to the challenge. "I did everything I could," Skelton said. "I felt it was an opportunity to try to convince them that a woman could do this type of thing and could do it well."[21]

Betty Skelton was a charismatic choice for the feature. Vivacious and daring, the five-foot, three-inch, thirty-three-year-old Skelton was also a serious pilot known especially for an acrobatic maneuver called the "inverted cut ribbon." While assistants held a ribbon suspended ten feet off the ground, Skelton would slice it in two with the propeller of her plane while flying upside down. Holder of two world altitude aviation records, she also loved skydiving, riding motorcycles, boat jumping, and racecar driving, and four times set the women's world record for land speed. Skelton was the first woman to drive over three hundred miles an hour and epitomized what *Look* editors meant when they argued that compared to men "women are better power packages."[22] She was also a good sport, cheerfully interacting with the male astronauts and laughing along with jokes and the occasional putdown by aerospace officials. Talking about the prospect of women astronauts, one man told her, "If I had my way, I'd send them all out there." Skelton said, "[T]hat kind of gave me an idea of what was going on."[23]

The *Look* feature, competing with *Life* magazine, which had exclusive access to the Project Mercury astronauts, presented an inside look at what astronauts would wear, inhabit, and be faced with in outer space. Over a four-month period in late 1959 and early 1960, Betty Skelton traversed the country, participating with Scott Carpenter, Wally Schirra, Alan Shepard, John Glenn, and Gus Grissom in various astronaut exercises. She took the tilt-table test at the School of Aviation Medicine in San Antonio, tried on a space suit at McDonnell Aircraft in St. Louis, surveyed the world's largest centrifuge at the Naval Air Development Center in Johnsville, Pennsylvania, and inspected the Mercury capsule at Langley Field in Virginia. Skelton practiced weightlessness maneuvers in a swimming pool with an Aqua-Lung—a feat that tested her will since she did not know how to swim.[24] Skelton found Dr. William Douglas, the astronauts' physician, "a little less negative than most" of the men affiliated with the space program because he believed women might in fact be better equipped than men when it came to handling periods of inactivity in space. Skelton assumed that Douglas found women uniquely suited to cope with tedium, since so many women were housewives who did the "same things over and over and over again" with little intellectual stimulation to engage their minds.[25]

Though the *Look* article stated that "no exclusive data" existed on how women compared to men in facing the physical and psychological stresses of spaceflight, Skelton's test results were strikingly absent from the story. Rather, Skelton was depicted listening to and receiving guidance from the astronauts. She was photographed in men's pajamas, her hands, playfully thrust into pockets on either side, extending the large folds of extra material as a child would in dramatizing "grownups'" clothes. Technicians at Brooks Air Force Base had handed Skelton the pajamas when they confessed that they did not stock suitable clothing for women who might undergo testing. "I think that just only goes to point that they really weren't thinking about women very seriously and did not have the equipment there to even test women at the time," she said.[26] While Skelton's abilities did not appear to have been overtly questioned by the astronauts themselves or any of the space officials she encountered, it was clear that her experience was nothing more than a staged publicity stunt.

After talking to physicians, military men, aerospace engineers, and psychologists, *Look* summarized the characteristics of the ideal first woman in space and offered a thumbnail sketch of the future female astronaut. She would be younger than thirty-five, married, a pilot, and an outstanding athlete who excelled in swimming and skiing rather than masculine sports such as wrestling. Her scientific knowledge would span "astronomy to zoology" and she would be brought aboard perhaps as a "scientist-wife or a pilot-engineer." Foreseeing inevitable complications with the fit of space suits designed for the regulation male, the ideal woman astronaut would be "flat-chested" and not "bosomy." Her ability to provide social comfort and stimulation to others on the crew would be important as would be the capacity to cope with isolation.[27]

One of the scientists *Look* interviewed for the article was Don Flickinger. He offered insight for the article, was photographed with Skelton, and forecast that while women would not be used immediately as astronauts, they stood a chance of being included when semimaneuverable three-person orbital spacecraft came into use.[28] By the time he consented to participate in the Betty Skelton feature, it was likely that Flickinger already knew that the Wright-Patterson "girl astronaut" program had been scrubbed. Nothing was mentioned in the article of his experimental program or its rapid demise. In fact, although it was an article that claimed to provide a glimpse at how Project Mercury astronauts were tested, both the Lovelace Foundation and the Aerospace Medical Division were conspicuously absent from the story. No doubt Wright-Patterson and Colonel John Stapp would not give approval for any more women astronaut testing, be it official or staged. Randy Lovelace, of course, was preparing for Jerrie Cobb's tests in Albuquerque and even had another magazine spread of his own in the works. As staff at the Lovelace Foundation readied for a woman to take the Project Mercury astronaut tests, Ralph Crane, a photographer for *Life* magazine, prepared to photograph Cobb's every step for an upcoming exclusive.

Driving down on Gibson Avenue from the Albuquerque Airport, the Lovelace Foundation and Clinic buildings were easy to spot. Designed by John Gaw Meem, whose La Fonda Hotel was one of Santa Fe's most photographed structures, the medical building was a stunning example of pueblo-inspired architecture. As visually striking as Meem's design was, it also brilliantly depicted the institution's founders and their twin aims: the clinic's solid commitment to patient care and the foundation's visionary research. With the Sandia Crest Mountains behind them, the whole scene turned pink—a kind of luscious watermelon hue—for a short period of time every night at sunset. At precisely that time of evening, the Lovelace complex looked like four giant silhouetted adobe steps that climbed toward the distant mountains and into the sky.

When Lovelace patients and physicians found it too confusing to distinguish William Randolph Lovelace from his nephew William Randolph Lovelace II, the two men became simply Uncle Doc and Randy. They became as inextricably connected to the institution that bore their names as the after-hours telephone that rang simultaneously at the hospital switchboard and Uncle Doc's house. Like so many "lungers" who settled in the Southwest for the clean air and relief from respiratory ailments, Uncle Doc moved, in 1906, to Sunnyside, New Mexico (later Fort Sumner), for his health and a chance to use his new St. Louis medical degree, working as a physician for the Santa Fe Railroad. When Uncle Doc suffered a medical relapse, most of his family, including his young nephew, left their native Missouri and traveled to New Mexico. The Lovelaces stayed on and Uncle Doc recovered, making medical calls on horseback with a $2.50 Iver Johnson revolver in his bag.[29] After his parents' divorce, Randy Lovelace began spending more and more time with Uncle Doc, who had moved to Albuquerque, where he was building a thriving group medical practice. Randy Lovelace returned to Missouri for his undergraduate education at Washington University, then studied briefly at Cornell before taking his medical degree at Harvard in 1934. An internship followed at Bellevue in New York, where he hoped to stay until Uncle Doc persuaded him that the Mayo Clinic—his uncle's ideal of a vibrant medical facility—was worth a look. Lovelace agreed and began a fellowship in surgery. Work with his mentor, Dr. Walter Boothby, further increased Randy Lovelace's fascination with aviation medicine and led to their collaboration with Dr. Arthur Bulbulian on the BLB high-altitude mask.

By the late 1930s, Randy Lovelace had settled into life in Rochester, Minnesota. He and Mary Moulton of Albuquerque, whom he had married in 1933, were busy raising three children: Mary Christine, born in 1938; William Randolph Lovelace III, born in 1940; and Charles Moulton, who followed in 1942. Lovelace took a leave from the Mayo Clinic in 1942 and, with his credentials as flight surgeon, served in the Army Air Force—time marked by his famous parachute jump and a stint as chief of the Aeromedical Laboratory at Wright Field. After the war, Lovelace returned to his career at Mayo while the Lovelace Clinic, under Uncle Doc's direction, grew to be the top facility for patient care in Albuquerque.

Randy and Mary Lovelace might well have centered their lives on their growing family, his work at the Mayo Clinic, and her performances as a violinist with the Rochester Symphony. In 1946, however, family tragedy struck. A polio epidemic swept through Minnesota during the summer of 1946, and young Ranny died suddenly on July 7. As his parents prepared for their son's burial in Albuquerque, Charles came down with the same devastating symptoms as his brother. Lovelace's good friend Jackie Cochran immediately flew to New Mexico to help. In her private plane, Cochran took Mary Lovelace and three-year-old Charles on board and piloted them to Warm Springs, Georgia, to President Roosevelt's famed polio hospital. Several weeks later, Charles died, and Mary and Randy Lovelace buried their second son.

The couple simply did not have the heart to return to Minnesota. They stayed in New Mexico among family and took solace in the wild, open country that they both loved. Two daughters were born soon afterward: Sharon in 1947 and, two years later, Jacqueline, named for her godmother, Jackie Cochran. Soon Uncle Doc approached Randy Lovelace with an idea about expanding the clinic. Would he be interested in remaining in Albuquerque and starting a new research foundation—a facility that, in part, would focus on aviation medicine? Lovelace accepted his uncle's proposal, and the men went to work. By 1947, using the Lovelace Clinic as a base, the Lovelace Foundation for Medical Education and Research was born. Floyd Odlum, Jackie Cochran's wealthy financier husband, became chairman of the foundation's board of trustees. With a deep personal interest in aviation, Odlum used his money and his extensive influence to support Lovelace's bold experiments. The foundation rapidly expanded with research projects for the National Institutes of Health, the Atomic Energy Commission, the Air Force, and NASA. With Uncle Doc overseeing the clinic and Randy Lovelace heading the foundation, the Lovelace institutions thrived. Many Air Force physicians went to work for the Lovelace Foundation after they retired from the military. Rich collaborations developed among Lovelace, the Air Force School of Aviation Medicine, and the U.S. Air Force Air Research and Development Command. When Jerrie Cobb walked into the lobby of John Gaw Meem's great pueblo-style medical building to begin astronaut tests in February 1960, she scarcely could have imagined the empty mesa that had occupied the space a decade earlier. Now, with the impressive Lovelace medical building squarely in place, the landscape on Gibson Avenue absolutely pulsed with ambition.[30]

Cobb arrived at the Bird of Paradise Motel in Albuquerque on Sunday evening, Valentine's Day. As luck would have it, Randy Lovelace was out of town for the week. His absence was not unusual. Doctors at the Lovelace Foundation and Clinic were familiar with his hectic pace. "Many's the day Randy would do a couple of gallbladders in the morning and have a plane to catch at 12:30 and be on it," Don Kilgore said.[31] Other doctors simply picked up the slack or Lovelace appointed someone to stand in his stead. Therefore, Dr. Robert Secrest, an internist, telephoned Cobb at the motel that evening, greeted her, and inquired if she had any questions before beginning the next morning. Familiar with the Project Mercury men's military adherence to "need to know" policy, Cobb did not ask for any additional information. She chatted briefly with Secrest and then retired, making sure she followed the doctor's directive of not eating or drinking anything after midnight.

The next morning, twenty-eight-year-old Jerrie Cobb crossed Gibson Avenue and walked into the Lassetter Laboratory Building to begin astronaut tests. Passing through the door bore a different significance for Cobb than it had for the Project Mercury candidates who had made the same walk a year before. Carpenter, Cooper, Glenn, Grissom, Slayton, Shepard, Schirra, and twenty-four other male candidates had been competing against one another for a chance to be selected as one of the nation's first astronauts. Their entrance had been part of an epic journey that landed seven of them in space capsules and created a thrilling human milestone. The federal government had already assumed that several of the men would get the chance to be launched into space. Had Shepard or Glenn not succeeded, other men would have been selected in their place. If Jerrie Cobb did not succeed, however, neither she nor any other woman would likely get another shot. The government, as Ruth Nichols had pointed out, controlled who went through the door to space. If Cobb failed, Flickinger's and Lovelace's experiment would have been just that—an experiment with a negative result that confirmed many assumptions about women's limited physical ability.

Over the next six days Cobb would battle tilt tables, electrical stimulation to the nerves, three feet of rubber hose slithered down her throat, exhausting physical endurance exercises, frozen ears, frozen hands, frozen feet, brain-wave measurements, radiation counts, and the nightly dose of humility—barium enemas. In all, doctors scheduled a total of seventy-five tests to measure the range of her body's capability. Yet no battle was more formidable than the one she faced outside the Lovelace doors: the battle against sexism. Like Ruth Nichols and Betty Skelton before her, Cobb understood that if she measured up, people who resisted the idea of women in space might be forced to rethink their position. Men such as Colonel John Stapp, who had voiced his opposition to testing women at Wright-Patterson, would have less credibility when asserting that women lacked strength, endurance, will, and even emotional stability. There was more than just her own success on the line for Jerrie Cobb at eight o'clock that Monday morning. At stake was the chance to open doors on a permanent basis for any woman who wanted to fly a jet or guide a space capsule.

Cobb's chance to prove herself began that day with a trip to the lab and a stool specimen. Blood counts, serology, liver function tests, protein-bound iodine, blood sugar, cholesterol, Rh factor, sedimentation rate, urinalysis, medical history, and an EKG all came before lunch. After her meal, Cobb spent the afternoon undergoing a proctoscopic exam and finished the day with vector cardiograms and X rays of her sinuses and teeth. The tests were identical to the ones for Project Mercury astronauts established by the NASA Special Committee on Life Sciences. Lovelace instructed his physicians to apply the same scrutiny to Jerrie Cobb as they did the male candidates. Nor were the staff to limit their expectations for Cobb because of personal notions of women's capabilities or their own sense of chivalry. Cobb should be pushed as far as she could go.

Assembling comparative data was the goal of the Lovelace clinical examinations. In testing the male astronauts, NASA wanted to calculate how each man compared to another, so physicians determined what NASA called "degrees of physical soundness . . . and evaluation was dependent upon a comparison of each man to his fellow candidates."[32] At each stage of the exam, candidates were judged not on the basis of merely passing or failing a specific test but rather on how their scores stacked against each other. At the end of the week, Cobb's scores would be compared both to those of the thirty-one jet test pilots who had been tested at Lovelace the previous year and to those of the seven men whose scores were at the top of that list and who eventually had been selected as astronauts.

As much as physicians at Lovelace tried to objectively compare Cobb's data to the male candidates', subtle hints of gender bias occasionally revealed themselves in their language and, inevitably, in their thinking. Dr.

Secrest, for example, a man Cobb found "earnest and compassionate," nevertheless indicated that he regarded men as the physiological standard and women as variants. "There was the problem that women have physiological variation, and are not always physiologically the same as men," he said. "So we got some good minds in the GYN department to help us along [in devising testing criteria for women] . . . if you did a set of tests on a woman, then during her menstrual [period] you would have to do them over again to see if she'd react the same way. So it was a little more difficult. Women do vary physiologically from men, there's no question about it."[33] Although Jerrie Cobb's statistical data might reveal that she and the male candidates had performed equally well on the examinations, cultural attitudes toward women that viewed her as a "variation," her physiology as a "problem," and the creation of her tests as "more difficult" suggested that Secrest viewed Cobb as a kind of mutation: an alteration of the male norm. Such bias, while nearly invisible, had profound effects nevertheless. Like medical research that tested only white male heart patients and then applied the results to men and women of all races, the practice of seeing white men as the medical norm was widely accepted and rarely challenged. It viewed women as puzzling versions of men. Or it simply did not see them at all.

Jerrie Cobb continued with her testing throughout the week. Tuesday she took the "Puritan ordeal" exam, which used water displacement for measuring her total body fat. Lovelace wrote that "density of the whole body was determined by weighing the nude body in air and when completely submerged in water with the lungs emptied."[34] Subjects were suspended from a ceiling beam while seated in a chair and then lowered into the water. In the afternoon, Cobb reported to the clinic wing for the tilt-table test and an evaluation of her cardiovascular system. She boosted herself up onto a long, hard plank, and technicians wired her body for electrocardiograms and blood pressure readings. Stretching out, she tucked her hands behind her waist and looked beyond her plaid shorts and moccasins to the green wall at the end of the room. Then the wall started to move, or so it seemed to Cobb. From her horizontal position, she was tilted to a steep sixty-five-degree angle and back down again to horizontal. For thirty minutes Cobb stared straight ahead while doctors recorded data about her blood pressure and heart rate every sixty seconds. She did not become dizzy, although she did worry about slipping off the table since there were no straps to hold her on.[35]

More tests followed: gastric analyses, colon exams, lumbar-spine X rays, ophthalmological exams for visual fields, tonometry, slit lamp studies, depth perception, night vision. Lovelace reported that the analysis of the eye exams followed the strict U.S. Air Force standards used to evaluate pilots.[36] Additional tests included pulmonary function trials, bicycle ergometer pedaling to calculate physical competence, and blood volume determinations conducted by inhaling a small amount of carbon monoxide and tracing the amount absorbed in the blood. For the test called the Valsalva maneuver, Cobb blew as forcefully as she could with her nose and mouth closed against a 50 mm mercury column for fifteen seconds while doctors read her blood pressure and listened to her heartbeat for any cardiac defects. And with some reluctance, Cobb submitted to doing Grandfather's banana oil recitation. On the taped replay, she found her voice slightly unsteady but otherwise fine.[37]

One "test" that was not on the official schedule but that was deemed important for social reasons was a dinner party. Since Randy Lovelace was out of town for the week of Cobb's tests, Dr. A. H. Schwichtenberg stepped in and invited her for an evening at his home. Cobb would miss meeting Lovelace's wife, Mary, a woman whose graciousness hardly suggested that she had been born in a tent in New Mexico and killed a few snakes in the process of planting grass on the family's arid Albuquerque lawn.[38] Cobb also would not hear Randy Lovelace recount the story of the impressive paneling in his den—wood he had salvaged from the White House during renovations during the Truman administration.[39] To Jerrie Cobb, however, the evening with Dr. Schwichtenberg was certainly enjoyable. Schwichtenberg, a retired Air Force brigadier general, had developed NASA's testing protocols for astronaut selection. He went to Lovelace after his career in the military to head the Department of Aerospace Medicine and Bioastronautics. Cobb was thrilled when she learned that also joining General and Mrs. Schwichtenberg for dinner would be aviation legend Jimmy Doolittle and his wife, Jo.[40] Cobb said the Doolittles "were among the most 'comfortable' people I've ever met."[41] Certainly General Schwichtenberg's enthusiasm for space and women astronauts also made the evening a success. Like Randy Lovelace, Schwichtenberg had been thinking about space exploration many years before it entered other people's minds. "My wife reminds me that I told her I expected to see us put a man on the moon while we were still courting," he remarked. "[A]nd we got married in 1929!"[42] He also cared deeply about testing women astro-

naut candidates. "Randy and I had talked about this a number of times—
what about women in space?" he said. He believed women and men were
equal on all counts. Cobb's exam results that week were already underscor-
ing his point. "That proves one of Schwichtenberg's laws: that there's just as
many smart women as smart men!" he said. "The second one is, there are
just as many dumb women as there are dumb men."[43]

On Thursday morning, Cobb's testing schedule changed slightly with a
top secret flight to nearby Los Alamos for what physicians called a "total
body count"—an evaluation of Cobb's body for radiation and lean body
mass as determined by potassium calculations. Amid the tight security at the
Los Alamos Scientific Laboratory, Cobb met Dr. Wright Haskell Langham,
who served with Randy Lovelace on NASA's Special Committee on Life
Sciences and, even before the creation of NASA, had worked with Flickinger,
Scott Crossfield, and Lovelace on the Human Factors ad hoc group for the
Special Committee on Space Technology. Cobb's afternoon at Los Alamos
was memorable more for its secrecy and strangeness than for any arduous
physical tests. She recalled descending two flights of stairs from Langham's
office to a basement room dominated by a huge silver cylinder, an eerie
machine that looked like an early model of a magnetic resonance imaging
(MRI) machine. Technicians explained that she would slide into a smaller
tube that would then move into the larger cylinder. Once inside, she would lie
still as the machine read her radiation count and potassium level. She
climbed in and took the "chicken switch" that assistants handed her just in
case the confined tube engulfed her with claustrophobia. As she glided forward into the dark, cool cylinder with her arms folded across her chest, Cobb
had the sensation of being inserted into an iron lung. She carefully reached
up, trying to feel how much room she had to move. Eight inches. Panic did
not set in. Cobb waited patiently, holding but not triggering the chicken
switch, until the counter whirred into motion again and spit her out into the
stark light of the basement room.[44]

The most painful test of all that week was Dr. Donald Kilgore's ear exam,
which measured Cobb's equilibrium and response to dizziness. Nearly every
Mercury 7 astronaut candidate found the test uncomfortable, if not outright
painful.[45] The test began with Dr. Kilgore turning Cobb's head so that one ear
faced the ceiling. He then took a large syringe, what Cobb remembered as
one "so big it looked like a teaching model," and inserted it into her right

ear.[46] Slowly cold water at ten degrees Centigrade began trickling against her eardrum in an experiment designed to make her balance go haywire. Cobb cringed; the lights around her swirled. She tried reaching for the arm rest of her chair but could not make her hands work. As a nurse stood by, timing her reactions with a stopwatch, Cobb's eyes began to show the expected signs of nystagmus—her eyeballs rapidly oscillated in an involuntary motion. "Precise measurement of how soon nystagmus began, how violent it was, and how it lasted gave us an idea about how the pilot might fare when orbiting in a gravity-free state," Lovelace wrote.[47] For Cobb, like nearly everyone else, the worst part of the test was not the initial onset of staggering disorientation but the excruciating sensation of watching the syringe come toward her a second time and knowing what to expect.

Cobb's final appointment at the Lovelace Foundation came on Saturday morning when she met with Dr. Secrest for a preliminary review of her scores. While a complete analysis would come later, after Dr. Lovelace and others had had an opportunity to compare her scores with the Project Mercury astronauts, initial indicators suggested that she had done very well. Cobb's only notable deficiencies were a slight hearing loss in her left ear—not uncommon for pilots of multiengine planes who sit near loud engines—and cold feet.

Cobb's experience at the Lovelace Foundation galvanized her ambition, cold feet notwithstanding. She searched for opportunities, like flying a jet when she could, that would add more credibility to her potential astronaut candidacy. Even though she knew that Randy Lovelace was testing her out of his own curiosity and did not have NASA's sanction, Cobb began to take herself seriously as a candidate. She wrote letters to Wright-Patterson Air Force Base requesting permission to undergo stress tests. She was turned down. Even more discouraging was the fact that the reply came from the office of public information and not the Aerospace Medical Division. Clearly the Air Force perceived her as a public relations issue, not an astronaut candidate. "Because the equipment is in use and there exists no requirement for stress data on female tolerance, we regret to inform you that we are unable to comply with your request," the letter read.[48] The spirit of cooperation at Wright-Patterson that led to tests for civilians such as Ruth Nichols was no longer evident for women like Jerrie Cobb.

Not all the answers Cobb received were disheartening, however. NASA's own Lewis Research Center in Cleveland, for example, approved her request to undergo spaceflight simulation testing on its special equipment. While not an official astronaut candidate, Cobb was afforded unique treatment reserved for guests with influential connections or those whose aviation abilities were unquestioned. During its early years, such impromptu decisions by NASA were not uncommon and often reflected one individual's judgment rather than NASA's official position. Ad hoc decisions were possible before the agency grew into a vast hierarchy with complicated chains of command. The open door Cobb walked through at the Lewis Research Center, as with the Lovelace Foundation, was available to her because a single person or a group of individuals was convinced that testing her was a good idea. Cobb took every advantage of those opportunities.

At Lewis, Cobb wanted to ride the MASTIF, the Multi-Axis Space Test Inertia Facility. It provided one wild ride. Sometimes called a gimbal rig, the MASTIF was an enormous gyroscope, some nineteen feet across, that spun its rider in three directions at once, attempting to simulate what scientists believed an out-of-control space capsule might feel like. John Glenn wrote that for testing the limits of one's control, the MASTIF was "the diabolically perfect machine."[49] In one direction, the MASTIF "pitched," moving wildly up and down. In another, it "rolled," slanting sideways. When it "yawed," it spun in circles. Even if one had the stomach for carnival Tilt-A-Whirl rides, the MASTIF was in a nausea-producing category all by itself. Project Mercury astronauts "graduated" from MASTIF testing when they were able to bring the rig under control while doing thirty revolutions per minute on all three axes.[50]

Jerrie Cobb spent forty-five minutes on the MASTIF and maneuvered the two hand controls to bring the three whirling motions to a halt. It was no easy task. "First the thing started to pitch," she wrote in her autobiography, "and if I hadn't been fastened in, I would have been tossed right off the couch. Then as the pitch reached peak speed, I felt the roll start. I was twisting, twisting like a top, and going head-over-heels at the same time. As yaw set in, I realized the cage and everything in my line of vision was a dizzying blur."[51] NASA photographers, perhaps at Cobb's own insistence, photographed her ride, catching her in the center of the gyroscope with both hands gripping the controls and surrounded by crisscrossed girders and wires. She looked like prey caught in a steel trap, except that there was not a trace of fear, only intense concentration, as she stared straight ahead.

Ivy Coffey did not even know her housemate had disappeared for astronaut testing until Cobb returned home and told her about it. It was common for the two women not to see each other for stretches of time since they both were on the road so much. Cobb did tell Coffey some of the details of her testing, however, and Coffey got used to the occasional phone call from Randy Lovelace, often late at night, trying to reach Cobb. As the time neared for publicly releasing news of her testing, it became more and more critical that Lovelace and Cobb confer. A carefully orchestrated chain of events needed to occur once Lovelace made his announcement at the upcoming international space medicine symposium in Stockholm. Cobb would then hide from reporters until *Life* magazine published its exclusive of photographs taken in Albuquerque. Only then would she emerge for a fascinated public.

By August the time for the announcement had come and Lovelace was thrilled with Cobb's results. "There is no question but that women will eventually participate in space flight therefore we must have data on them comparable to what we have obtained on men" he said in a note that accompanied his handwritten graph detailing Cobb's scores on thirteen laboratory tests. "For example you will note that Jerrie requires less oxygen per minute than the average male astronaut—this means less oxygen by *weight* will need to be carried for the women crew members than for the men. We will ultimately examine 12 or more women pilots."[52] With additional test scores, assuming they were as first-rate as Cobb's, Lovelace could marshal his case for women astronauts and expand his one-man experiment. Lovelace's scientific paper, authored with Dr. Schwichtenberg, Dr. Secrest, and Lovelace Foundation pulmonary specialist Dr. Ulrich Luft, was finished after many drafts. Cobb was in New York introducing Aero Design's new aircraft models with the understanding that once the news of her testing broke in Stockholm, she would vanish until *Life* hit the newsstands. Lovelace boarded a plane for Sweden.[53]

AT 3:34 A.M. ON FRIDAY, AUGUST 19, THE TELEPHONE RANG IN THE undisclosed New York apartment where Jerrie Cobb was staying. The telephone rang at Cobb's parents' home in Ponca City, too. Reporters from England, Japan, Australia, and all over the United States had begun a frantic search for the girl who—they had just heard—had passed the astronaut tests. Cobb's friends who were contacted by the press played dumb. Aero Design did not divulge her whereabouts. *Time* magazine wanted an interview with Cobb and strong-armed the editors at *Life*, their sister publication, for a portion of their exclusive.[54] By morning Cobb's photograph was in newspapers across the country. She looked radiant in the publicity shot, smiling from the cockpit of a plane, coiffed and wearing a string of pearls. *The New York Times* reported that Dr. Randolph Lovelace had disclosed in Stockholm that Jerrie Cobb, a twenty-nine-year-old Oklahoma City woman, had "successfully completed the tests given to the seven men in the United States men-in-space project." Lovelace, whom the newspaper identified as the "top United States space medicine expert connected with the Mercury Project," conceded that the first flight of a woman astronaut was a long way off and that "there was no definite space project for the women."[55]

Back home in Oklahoma City, Ivy Coffey's editors at *The Daily Oklahoman* wondered why their newspaper had not scooped the world with news of Jerrie Cobb's testing, since, after all, one of their writers shared a home with the world's current celebrity. Coffey, who often had found her male editors not as interested as she hoped they would be in stories about the accomplishments of a local girl pilot, swallowed her "I told you so" and set to work cranking out a story with the local angle for the Friday edition of the newspaper. Even with a news blackout in effect, Coffey was able to snag at least one quote from Cobb that no one else had. "This isn't negative thinking," Cobb told Coffey, "but I'd want to [travel into space] even if I didn't come back."[56]

By Tuesday morning a full-court press was on with the nation's media. *Life* would be published in a little over a week, and a *Sports Illustrated* article by a journalist named Jane Rieker, whom Cobb had met in Cuba in the 1950s, would follow, emphasizing Cobb's "can-do" attitude.[57] Editors at *Life* decided it was time to bring Cobb forward, and she faced an eager press corps at the Time-Life Auditorium in downtown Manhattan. All the anxiety Cobb had felt in the past about answering reporters' questions was nothing compared to what she now faced. The questions were unceasing and at times ridiculous. "Can you cook?" one reporter asked. Cobb remembered one of Ivy Coffey's specialties and said that Chickasaw Indian dishes, particularly smothered steak, were her favorite. To get through the surreal crush of publicity, another side of Cobb's personality had to take over. The public persona pushed forward and left shy Cobb in the wings. As Cobb later put it, "Jerrie and her alter ego, Geraldyn, had words."[58]

BLOCKS AWAY FROM JERRIE COBB'S FIRST PRESS CONFERENCE AS "THE lady astronaut," Ruth Nichols sat in her Manhattan apartment, discouraged and worried about her future. The family's fortune had long ago run out. The 1929 stock market crash and her father's final illness had forced the once affluent Nichols family to move into increasingly smaller homes. Now, at age fifty-nine, Nichols felt the pressure of earning her own way and helping the older members of her family. Few airlines would even think of hiring a woman, and charter work was also difficult to come by, especially when a pilot was over fifty. She found work where she could, serving as a field director first for the National Nephrosis Foundation and later the National Association for Retarded Children. She then signed on with Friendly Homes, a medical organization. While she was able to do some promotional flying for those organizations, the work was not what she had hoped. Tutoring herself as much as she could about space medicine, Nichols still dreamed of space-flight. "I'm going to fly until my last breath," she said. "Either in my present bodily form or another, when space ships take off, I'll be flying them."[59]

On Sunday, September 25, Nichols's Aunt Polly telephoned her niece repeatedly but no one answered. Finally, at two in the afternoon, she asked the building superintendent to unlock the door of Nichols's apartment and see if she was all right. The superintendent found Nichols dead on the bathroom floor. Later, the city medical examiner reported the presence of barbiturates in Nichols's system and ruled her death a suicide.[60]*

---

*Two of Ruth Nichols's friends believed her disappointment over not becoming an astronaut may have contributed to her suicide. Jane Hyde Fawcett wrote, "If anyone had talked to her, they would have known that the future for Ruth was still space. She hoped, she built her physical being into the future of a moon shot." Fawcett's friend Alice Benson also pointed to Nichols's space aspirations. "I suppose we will never even get a hint—whether yet another job had blown up—she had been turned down for space flying."[61]

CHAPTER FIVE

The List

RANDY LOVELACE KNEW THAT NASA WOULD VIEW JERRIE COBB'S exceptional test scores as a fluke and not representative of women pilots in general. He reviewed the list of names that Flickinger and Cobb had compiled a year earlier when Wright-Patterson tests still seemed possible. None of the women had ever been contacted. Lovelace studied the list and, with Cobb's help, vetted the existing names and suggested others. He hoped that at least two dozen women could be found who had the necessary aviation credentials and who would be interested in traveling to Albuquerque for the secret tests.

Jerrie Cobb threw herself into the task of assembling the list. She was in a much better position than Randy Lovelace to identify which women pilots might be considered potential candidates. Apart from his friendship with Jackie Cochran, Lovelace did not come into regular contact with many women pilots. And asking Cochran to help with the early selection process seemed ill advised. Cochran was just beginning an unprecedented second term as president of the Fédération Aéronautique Internationale. She was traveling abroad a great deal and was very occupied with her own work.[1] Fortunately, Cobb was acquainted with most of the women in the higher circles of American aviation at the time. Some, such as Jerri Sloan, were personal friends whom she knew well. Cobb met many others at Powder Puff Derby races and closely followed news about any women going after world records. She also became familiar with a large number of women pilots through her marketing work for Aero Design. Engulfed with his own demanding work at the foundation, Lovelace accepted Cobb's offer to compile a list and submit it for his consideration.

Between September 1960 and August 1961, the secret list began to take shape. Cobb would identify a pilot and forward her name to Lovelace, who in turn sent potential volunteers a questionnaire and introductory letter. While she assembled the list, Cobb continued to work in Oklahoma City for Aero Design and Engineering. She was fortunate to have a supportive boss in Tom Harris, who allowed her considerable latitude in cooperating with the Lovelace Foundation. Harris was proud of Cobb and personally interested in her astronaut testing. He had been standing with her at the fateful meeting in Miami when Cobb met Don Flickinger and Randy Lovelace for the first time. Harris also realized that any public attention Cobb garnered through the Lovelace project would likely shine on his own company as well. By allowing Cobb the flexibility to devote herself to the task of list making and by keeping her paychecks coming, Tom Harris and Aero Design, in effect, provided the equivalent of financial underwriting for the preparatory stage of Randy Lovelace's women-in-space experiment.[2]

What, exactly, were Cobb and Lovelace looking for? First, Lovelace wanted women pilots who had racked up more than a thousand hours in the air,[3] not an easy task in the 1960s. Since women were not allowed to fly for the military and were not being hired by airlines, they had to build up hours in other ways. Many worked as flight instructors, since flying with a student pilot gave them time in the air that they did not have to finance personally. For women pilots who held down jobs outside aviation, it could take as long as five years to build up a thousand weekend flying hours. Second, Lovelace wanted candidates who were in their early thirties, although he was willing to consider exceptions if they seemed singularly focused. Lovelace did not specify any thoughts regarding a volunteer's marital status, although he knew the WASPs had often come under fire for losing too many pilots to matrimony. Stipulating only that all candidates be healthy, accomplished, and active, he also did not indicate any specific race restrictions or preference, although he asked candidates to indicate their "lineage." Finding women who had test pilot experience or college degrees in engineering—as the Project Mercury astronauts were required to have—would be difficult, if not impossible. He would have only Jackie Cochran to evaluate if jet test pilot experience were a requirement for women candidates.*

In surveying the names and qualifications of possible candidates, Jerrie Cobb also searched for qualities she considered critical. She knew from her own testing experience that physical fitness and a daring disposition would be valuable. Perhaps gutsy air-race competitors or women who flew heli-copters, navigated seaplanes, or doused forest fires from the air would also make ideal candidates. She looked for an independent personality or a woman who displayed gumption and self-confidence. She searched for women who faced the unknown, who were eager, motivated, and who would risk just about anything to be given a chance. Being willing to risk every-thing was the one quality Cobb knew was essential. All she had to do was find about two dozen women who were that determined.

For months, Cobb pored over records at the Oklahoma City headquar-ters of the Ninety-Nines, the international organization of women pilots founded by Amelia Earhart. She spent long hours huddled over file cards, cross-checking addresses and birth dates, verifying flying hours and aviation ratings. The work was tedious, all-consuming, and often made her eyes ache from the strain. Cobb also checked FAA records, luckily stored in a new facil-ity right in Oklahoma City. In 1960, official FAA statistics indicated that nearly ten thousand American women were involved in aviation—most still at the lowest rung of student pilot. Seven hundred eighty-two women held commercial pilot's licenses[4] and Cobb reviewed their records, searching for women with even higher ratings and increased flight time. Cobb became concerned that the list she was submitting to Randy Lovelace—names she was largely familiar with—did not include any minority women. African-American women pilots met with greater discrimination in aviation circles than did white women and often had difficulty gaining hundreds of hours in the air and advanced ratings. Cobb broadened her search to women pilots outside the United States and found a South Korean woman pilot who met

---

*The first woman to fly a jet was WASP Ann Baumgartner Carl, who flew an experimental version of the Bell YP-59A fighter in 1944. The plane was so experimental that when it was parked a fake propeller was attached to obscure the fact that it was a jet. In 1953, French pilot Jacqueline Auriol was the next woman to fly a jet, followed by Jacqueline Cochran.

the flying requirements. Her candidacy never went forward, as Lovelace told Cobb that all women being tested had to be U.S. citizens.[5]

Cobb's friend Jerri Sloan barely noticed the business envelope with "Lovelace Foundation for Medical Education and Research" as the return address when she retrieved the mail one afternoon in the fall of 1960. The thirty-year-old Sloan had had a frantic day running her aviation business, Air Services of Dallas. Sloan had pilots to schedule, aircraft that needed maintenance, contracts to pursue to expand her clientele, even late-night test flights that she had to pilot herself. After picking up her children from school, Sloan simply tossed the mail on the kitchen table and began sifting through the bills and assorted letters in front of her. When she stopped and studied the return address on the Lovelace envelope, she stared at it, perplexed and intrigued. Sloan knew exactly who Dr. Randolph Lovelace was. Every pilot did. Sloan ripped open the letter and could hardly believe what she saw. "These examination procedures take approximately one week and are done on a purely voluntary basis," the letter read. This experiment must have been what Jerrie Cobb had been hinting at earlier, Sloan thought. Months earlier, the two friends had talked and Cobb had asked obliquely if Sloan might be able to get away for a few weeks for top secret government business. Sloan had not known what Cobb was talking about and chalked it up to her friend's mysterious ways. Sloan continued to read the Lovelace letter, her excitement mounting. These tests "do not commit you in any further part in the Women-in-Space Program unless you desire." By that point, Sloan had already made up her mind. She could not imagine any woman turning down the chance to participate in a women-in-space experiment, particularly one with her credentials: 1,200 flying hours, commercial pilot's license, multiengine rating, air-race honors, and experience flying B-25s on harrowing runs for Air Services to test infrared surveillance equipment. When she finished the letter and placed it back on the kitchen table, Sloan's nine-year-old son, David, looked at his mother's astonished expression and begged her to read the letter out loud. Sloan started at the beginning. "We have been informed that you may be interested in volunteering for the initial examinations for female astronaut candidates." That was all David had to hear. He burst out the door with a crash, whooping, waving his arms, and yelling to all his friends, "Mommy's going to the moon!"[6]

The neighbors, she remembered, did not think much of David's pronouncement; they already thought Jerri Sloan was crazy. After all, how many women in suburban America during the 1960s went off at odd hours of the day and night to fly planes to undisclosed locations for secret reasons? When she could not convince a bank to give her a loan to get Air Services off the ground, Sloan asked a business friend, Joe Truhill, to join her in the company just so that she could have a man's signature on the loan papers. It galled her to genuflect to the discriminating practices of the bank—but her end run generated the money she needed. Now working on a contract for Texas Instruments, Sloan was flying all over the state and the Gulf of Mexico testing top secret terrain-following radar (TFR), jobs that required her to fly low over water at night, with props churning the waves and no visible horizon reference. Because the new equipment had to be evaluated the moment engineers had it up and running, Sloan was on call twenty-four hours a day. Neighbors would peer out the windows and roll their eyes when they saw her backing out of her driveway at midnight.

For Jerri Sloan, going to the Lovelace Clinic was not only a rare aviation opportunity, it was also a chance to participate in a national cause—the space race. She always regretted being too young to join the WASPs. As a child growing up in Amarillo, she remembered watching smartly dressed WASPs step off the train as they traveled to Avenger Field in Sweetwater for basic training. Their confidence and patriotism had impressed her, and Sloan had longed for a chance to contribute to the nation. "My grandmother fought for the right to vote. My mother helped fight a war. My older stepsister was an Army nurse in an evacuation hospital. I grew up seeing women do everything," she later said.* Sloan immediately wrote to Randy Lovelace and told him she was indeed interested in volunteering.[7]

All across the country that winter and into the next spring, a select group of young women pilots began receiving letters from Dr. Randy Lovelace. In San Francisco, Nashville, Akron, Kansas City, Washington, D.C., Houston, Hollywood, and Oklahoma City, women carefully opened the auspicious-

---

*Sloan's stepsister served with the 56th Army Evacuation Unit in Anzio, Italy. Her mother, like many civilian women during World War II, had volunteered for Red Cross duty and trained as a nurse's aide. With many nurses serving in army hospitals, in VA hospitals, or in the Army nurses' corps, civilian women helped fill some of the gaps in stateside hospitals. The nursing shortage during World War II was also caused by segregation that restricted African-American nurses from caring for white soldiers.

looking envelope from the Lovelace Foundation and found themselves handed an opportunity they never could have imagined. For some of the women, such as Jan Dietrich, it seemed too good to be true. That was when her sister, Marion Dietrich, stepped in. The identical twins had learned to fly together as teenagers, and Jan was currently making a living as a flight instructor and a pilot for a construction company. Marion flew only on weekends or when her job as a reporter for the *Oakland Tribune* allowed. Marion was even more surprised when she found out that she, too, had received an invitation for astronaut testing. When she heard that Jan had cold feet about the Lovelace testing, Marion wrote her sister a quick and emphatic letter. "Your hesitating to go to Lovelace absolutely shocks me," she declared. "Jan, we are poised on the edge of the most exciting and important adventure man has ever known. Most must watch. A few are privileged to record. Only a handful may participate and feel above all others attuned with their time. To take part in this adventure, no matter how small, I consider the most important thing we have ever done. To be ASKED to participate, the greatest honor. To accept, an absolute duty. So, go Jan, go. And take your part, even as a statistic, in man's great adventure."[8] After reading Marion's letter, Jan Dietrich filled out her questionnaire as her sister had done, rented a stationary bicycle at a sporting goods store, and began training every night after work.[9]

A few women, such as twenty-one-year-old Wally Funk, were not on Jerrie Cobb's list and did not initially receive a letter from Randy Lovelace. Funk had read the article about Jerrie Cobb's testing in *Life* magazine and could not stop thinking about women in space. Funk, a Taos, New Mexico, native, was a flying instructor at Fort Sill, Oklahoma. She knew Cobb slightly from Ninety-Nines events around Oklahoma City, and when she read in *Life* that Dr. Lovelace would be testing additional women, Funk went straight to her typewriter. "I am most interested in these tests to become an Astronaut," she wrote. She then offered her credentials: university degree from Oklahoma State, collegiate flight team awards, flight instructor rating, single-engine seaplane rating, and 3,000 flying hours—over three times the number of hours Lovelace was seeking and an extraordinary amount for someone so young. Young Wally Funk's audacious request caught Randy Lovelace's attention, and he invited her to the foundation. Funk immediately began her own exercise program, pedaling her bicycle every day to Fort Sill and leaving her red '59 Vauxhall parked at home.[10]

As time went on, additional women came to Dr. Lovelace's attention by more idiosyncratic means and often without Cobb's knowledge or vetting.* Wally Funk ran into twenty-four-year-old Gene Nora Stumbough at an Oklahoma collegiate aviation gathering. Funk knew that Stumbough was a flight instructor at the University of Oklahoma, and told her about the Albuquerque tests. Just as Funk had done, Stumbough wrote Lovelace quickly, offered her credentials, and informed him that she simply could not imagine a women's astronaut testing program going forward without her participation.[11] Randy Lovelace agreed. Another candidate, B Steadman, the calm, self-reliant owner of the flight operation in Michigan, contacted Dr. Lovelace after she received an invitation and urged him to consider inviting her friend Janey Hart. Steadman and Hart were old friends from the Michigan chapter of the Ninety-Nines. She respected Hart's combination of daring and common sense, qualities that came in handy when Hart moved with her husband and eight children to Washington, D.C., in the late 1950s. Philip Hart had been elected U.S. senator from Michigan, and Janey Hart made it clear that she did not consider being a senator's wife her full-time occupation. She was a pilot, too. She flew helicopters and captained her own sailboat. Even though she was forty years old, the oldest woman to be invited for astronaut testing, Hart made the cut.[12]

Not every woman who received an invitation accepted the offer or could arrange for time off work. Some women believed they would be penalized for participating in "this Buck Rogers nonsense," and their jobs would not be waiting for them when they returned from Albuquerque.† Other women,

---

*Even forty years later, some of the Mercury 13 still do not know exactly how their names appeared on the list. Sarah Gorelick Ratley has never been sure how she was selected. Rhea Hurrle Allison Woltman believed that Jackie Cochran, whom she met briefly in Texas, might have suggested her. Woltman is careful to emphasize, however, that she has no hard evidence to support her hunch. Late in the selection process, Jackie Cochran began suggesting names to Randy Lovelace. Myrtle Cagle was the only candidate she referred to Lovelace who was asked to come to Albuquerque and who passed the tests. See Chapter 6.

†Dorothy Anderson's boss would not let her take time off work from the large flight school in Bluffton, Ohio, where she served as the only full-time flight instructor. "I would have loved to go," she said. Sylvia Roth, a corporate pilot for *Encyclopaedia Britannica*, also could not leave her job, and a colleague at work who claimed he had connections in Washington, D.C., told her that the chances of a women's astronaut program going forward were not good. Roth passed the word along to her friend Frances Miller, the operator of a fixed-based operation in Columbia, South Carolina, who also had been invited to the

such as Akron schoolteacher Jean Hixson, enjoyed unconditional support from their bosses. Hixson was reassured when her principal pledged not to tell anyone about her testing and offered to help personally in any way possible. "There will no leak from me, period," her principal vowed, and added, "Let's work on getting you in good physical condition."[13] Hixson had nothing to worry about as far as her physical conditioning and aviation credentials were concerned. They were impeccable and among the most outstanding of the group.

Hixson was the sole WASP among the candidates and had served under Jackie Cochran from December 1943 until the WASP deactivation a year later. While in the WASPs, Hixson had been stationed at Douglas Army Air Field and flew B-25s as an engineering test pilot. She also ferried planes, helped develop automated pilot measurement for T-31s, and measured weather conditions from the air. When the WASPs disbanded, the women pilots were given the option to join the Air Force Reserves as nonflying second lieutenants. Hixson did so and developed a close relationship with Wright-Patterson Air Force Base, where she later performed experimental research on the effects of zero gravity. In 1957, she was at the right place at the right time to snag a chance to break the sound barrier. When a newspaper reporter who was scheduled to break the barrier developed inner-ear problems, Hixson stepped in. She argued that her experience as a reservist and as an elementary school teacher put her in a good position to explain the effects of high altitude to students and the general public. The Air Force agreed, and Hixson, as a passenger, boarded a Starfire F-94 C jet with her school principal standing on the tarmac proudly snapping photographs. Hixson broke through the sound barrier at 840 miles an hour above Lake Erie. When she returned the next day to the Akron, Ohio, elementary classroom where she taught third grade, her students were awestruck. She became known as the "supersonic schoolmarm" and developed an aviation

Lovelace Foundation. Miller took Roth's advice and stayed home. Although interested in the idea of a women's testing program, Miller thought she would never be lucky enough to be chosen. Marilyn Link, whose name had been on the list Brigadier General Flickinger had passed on to Randy Lovelace, also turned down the invitation. Link assumed that since she was in her late thirties she was too old for any future astronaut program. Marjorie Dufton and Elaine Harrison also were invited, but records do not indicate whether they declined the opportunity or went to the foundation and did not pass the tests.

curriculum and a school planetarium and organized field trips to NASA's Lewis Research Center.[14]

Though her aviation background was superb, Hixson was deeply concerned about the one significant detail she lied about on her Lovelace questionnaire. Hixson looked at the long form requesting information about her height, weight, church affiliation, lineage, marital status, number of children, and surgical procedures. She carefully listed all her qualifications, including her WASP experience, four thousand–plus flying hours, high-altitude flying, explosive decompression experience, low-pressure chamber indoctrination, graduate degree in education, specialization in science and mathematics from the University of Akron, and her study of Russian as a foreign language. Hixson emphasized the value of a teacher in space in capturing the imagination of the nation's schoolchildren. Then she came to the line that read Age. Fearing that she would be disqualified for being too old, Hixson shaved two years off her birth date and reported that she was thirty-five. She would actually turn thirty-eight just a few days after she submitted the application. Unaware that Randy Lovelace already had invited forty-year-old Janey Hart, Hixson resorted to a desperate and uncharacteristic measure. Years later, after Hixson's death, her younger sister, Pauline Vincent, discovered the incorrect information among Hixson's meticulously organized carbon copies of correspondence. The falsified birth date made Hixson's sister smile. It was the only time she could remember Jean Hixson ever having fibbed.[15]

As reporters began asking questions about Jerrie Cobb's astronaut status and news leaked that Dr. Randy Lovelace was inviting additional women pilots to undergo testing, NASA felt forced to clarify matters. At a hastily called news conference, a NASA spokesman stated that the space agency "has never had a plan to put a woman into space, it doesn't have one today, and it doesn't expect to have any in the foreseeable future." Any news reports suggesting that a woman astronaut would soon be launched by NASA were simply untrue, he stated.[16] Despite NASA's protestations that it had not selected Jerrie Cobb, the media continued to view her as the country's first "lady astronaut."[17] It mattered little to the media or the public that it was not NASA but rather NASA Special Committee on Life Sciences members Dr. Randy Lovelace and Brigadier General Don Flickinger who had tapped a woman pilot for astronaut testing. Nor did it seem to matter that NASA was not paying for women to be tested at the Lovelace Foundation. Such distinctions, while accurate, were either unknown or blurred in the minds of most Americans.

Shrewd about public relations, Jerrie Cobb knew she must take advantage of the public's lack of discernment about who exactly selected her and who was selecting additional women candidates. She set out to present herself as a viable astronaut candidate, recognizing that an enthusiastic public might be able to persuade a reluctant NASA to consider instituting a women's program. The very media that Cobb had long regarded as an unwelcome irritation had now become an ally to be courted. She subscribed to a newspaper clipping service, followed press reports about herself, and took her case directly to the American people, speaking out with increasingly opinionated statements about the need for women astronauts in the United States' space program. As her list-making work for Lovelace neared completion, Cobb also assumed more of a leadership role among the women who agreed to be tested, answering their questions, suggesting ways they might prepare for the examinations, and keeping them updated. While fully aware that NASA had not accepted Dr. Lovelace's experiment as its own, Cobb was confident that everything about the testing was moving forward. Out of two dozen women who had been invited to the Lovelace Foundation, eighteen had accepted.* Now all that remained was for Dr. Lovelace to find time in the foundation's schedule for them to begin the examinations. Cobb was so consumed with compiling the list that she was taken completely by surprise when her leadership among the women candidates was abruptly challenged.

In the late fall of 1960, when Jackie Cochran learned of plans to test a group of women pilots for astronaut viability, she went directly to her friend Randy Lovelace and asked to know everything. Even though she was deeply immersed in her work as the first woman president of the Fédération Aéronautique Internationale, Cochran never wanted to miss a new challenge. She even admitted once that she had a "lifetime habit of insisting I

---

*Excluding Jerrie Cobb, the eighteen women pilots who were tested at the Lovelace Foundation were Rhea Hurrle Allison [Woltman], Frances Bera, Myrtle Cagle, Jan Dietrich, Marion Dietrich, Wally Funk, Sarah Gorelick [Ratley], Jane Hart, Jean Hixson, Virginia Holmes, Patricia Jetton, Irene Leverton, Georgiana McConnell, Joan Merriam [Smith], Betty Miller, Bernice "B" Steadman, Gene Nora Stumbough [Jessen], and Jerri Sloan [Truhill]. Bera, Holmes, Jetton, McConnell, Merriam [Smith], and Miller did not successfully pass the tests. Six other women were invited to undergo testing and declined participation. They were Dorothy Anderson, Marjorie Dufton, Elaine Harrison, Marilyn Link, Frances Miller, and Sylvia Roth.[18]

could do things I knew nothing about."[19] Cochran talked with Randy Lovelace and, by late November 1960, either accepted an appointment as his special consultant for women in space or simply appointed herself. She also offered some ideas of her own. Lovelace was not in a position to dismiss any help that Jackie Cochran might offer. She had been his friend for nearly twenty-five years, and her husband, Floyd Odlum, was serving as chairman of the Lovelace Foundation board of trustees. Odlum was an enthusiastic supporter of Lovelace's work with NASA and contributed generously to many of the foundation's innovative projects.

To Randy Lovelace, Jackie Cochran's contributions were hardly limited to her close friendship and the involvement of her wealthy and influential husband. He regarded Cochran quite simply as one of the most extraordinary American women of the twentieth century. Yet he must have been surprised when a four-page letter from Cochran landed on his desk outlining in roman numerals her specific suggestions for his women astronaut project. "As your special Consultant in connection with this program, I have a few tentative suggestions to make to you," she wrote. By the end of page four, there was little that felt "tentative" about Cochran's advice. She urged Lovelace to consider testing a larger group than the eighteen soon to be scheduled for examinations. Her experience leading the WASPs had led her to believe that many women would wash out or would not be able to pass the tests. Others would drop out for personal reasons, including marriage. She was concerned that fewer than twenty test subjects would not offer an adequate comparison with male astronaut candidates. Cochran also urged Lovelace to think twice about requiring candidates to have accumulated a thousand hours of flying—a requirement so difficult to achieve among women pilots that only the most mature women would be able to meet it. "When it comes to mental attitudes and emotional stability, you might not find it medically wise to have a group of unmarried oldsters on your hands," she warned. The best pilots in the WASPs, she argued, were women between the ages of twenty and twenty-three. "I don't think the difference between a few hundred and a few thousand hours of air time is going to make the slightest difference in the capacities of your candidates," she wrote. The fifty-four-year-old Cochran urged Lovelace to considered women who were years younger than the target age of thirty-five, but also women who were older. She concluded with a veiled criticism of Jerrie Cobb. All the attention being focused on one woman made her angry and envious. She made certain that Randy Lovelace understood that "there should be care taken to see that no one gets what might be considered priorities or publicity breaks."[20] Obviously she thought the spotlight was shining far too brightly on Jerrie Cobb.

Randy Lovelace incorporated some of Cochran's suggestions into his women-in-space program and rejected others. He was already open to accepting pilots as young as twenty-two-year-old Wally Funk and as old as forty-year-old Janey Hart. One suggestion that Lovelace readily accepted was Cochran's offer for her and Floyd to financially underwrite the testing program, and he later supplied her with a list of expenses for the women's travel, hotel, and meals while in Albuquerque. The Cochran-Odlum Foundation later gave the Lovelace Foundation stocks and certificates amounting to $18,700 to cover the cost of testing the women astronaut candidates.[21] Randy Lovelace did not ease the requirement of one thousand hours of flying time. He wanted accomplished and experienced fliers and was already keenly aware that NASA considered the women's lack of jet experience a problem. He also did not expand the number of women pilots for testing, although he asked Cochran for any additional names she might suggest. He did not want to waste time assembling hundreds of candidates. Already the women-in-space program had met with one setback when the Air Force had yanked General Flickinger's experiment. Lovelace wanted to get the testing under way immediately.

More conversations and meetings with Cochran took place at the Lovelace Foundation as the year drew to a close. Cochran even wanted to take some of the tests herself, and offered to tackle the bicycle endurance trial during one of her frequent visits to Albuquerque.[22] She wanted greater involvement with the experiment on every level, and asked Lovelace to copy her on his correspondence with the candidates. Cochran also began organizing files on each of the women, and paid particular attention to any public remarks that Jerrie Cobb was making. She made sure Lovelace saw any news reports that quoted Cobb as sounding as though she was the leader of the women's testing group. Cochran marked the passages she found objectionable and told Lovelace, "As you know, I think it important to have it a group operation with each participant individually anonymous until the whole group can be talked about without priorities or precedence."[23]

During one of her many speaking engagements that winter, Cochran roared into the Dallas airport in her Lockheed Lodestar. Standing on the tarmac as part of the official welcoming party were Jerrie Cobb and two women-in-space candidates, Pat Jetton, president of the Women's National Aeronautical Association, and Jerri Sloan, representing local Ninety-Nines. Cochran taxied to a stop in her plane as fans cheered and a news photographer attempted to take her picture. When the photographer asked Cochran to wave from the other window for a better shot, she would have nothing of it. "You're photographing from the copilot's seat," she yelled. "I'm the pilot." Jerrie Cobb must have felt that the press coverage that she had been cultivating so assiduously since Lovelace's Stockholm announcement was overshadowed by the headlong manner in which Jackie Cochran demanded attention. In another photograph that appeared later in Dallas newspapers, Cochran was front and center and Jerrie Cobb looked like a prop. It did not help Cobb's efforts to be taken seriously as an astronaut candidate to be referred to in the caption merely as a "fan" of Jackie Cochran.[24]

Jackie Cochran had two speaking engagements in Dallas, first addressing the Women's Group of the Dallas Council on World Affairs. Speaking to women's groups was not her favorite occupation. Chuck Yeager's wife, Glennis, remembered that "Jackie would get annoyed if any women's groups invited her to give a talk. 'What do I have in common with a bunch of damned housewives?' she would demand."[25] Jerri Sloan sat on the dais with Cochran for another talk, to business pilots, later that evening. Cobb did not attend. As Cochran began her speech, she painted a picture of her heroic work leading the WASPs. In an effort to praise the WASPs' risky flying, Cochran said, "The men were afraid to fly the B-17, and I said, 'My girls will fly them.' " Although Sloan never questioned the WASPs' courage, she was angry at the way Cochran hogged the credit and was embarrassed that Cochran had insulted male veterans in the audience. "These were good men," Jerri said, "and they were shocked by what she said." Sloan had never met Cochran before and did not defer to her as others frequently did. She had no appetite for Cochran's narcissism or her insults and thought her egotism reflected badly on all women pilots. "We got into it real big time," Sloan recalled. "Jackie," Sloan remembered saying, "I will tell you something. I will never sit on a dais with you again. Never. You embarrassed the hell out of me and I am ashamed to be a woman pilot." Cochran roared back at Sloan, "Damn you! And you want to be in the Mercury Program?" When Sloan responded that Cobb had already told her she was accepted, Cochran blew up. "Jerrie Cobb isn't running this program," she yelled. "I am!"[26]

CHAPTER SIX

The Bird of Paradise

NEVER ONE TO AVOID A DIRECT CONFRONTATION, JACKIE COCHRAN decided to talk with Jerrie Cobb about leadership of the women's astronaut program. She allowed a few days to pass after her heated exchange with Jerri Sloan, and approached Cobb with a cooler head and a measure of civility. Cochran wrote Cobb telling her she regretted that they "did not have an opportunity to get better acquainted" while they both were in Dallas. She then invited Cobb to the Cochran-Odlum Ranch in Indio, California—the first of many invitations for conversations on her turf. An invitation to Indio was more than a proposal for a genial weekend. It was a total immersion in Jackie Cochran's power and influence. The nine-hundred-acre ranch was opulent: lush grapefruit, tangerine, and date trees, a golf course, stable, heated swimming pool, and numerous guest cottages. Inside the main house was a magnificent dining room the size of a ballroom where generals, presidents, movie stars, and business tycoons lingered after cocktails and long dinners. On visits to the ranch, Cochran's namesake and godchild, Jackie Lovelace, used to steal away with her sisters to Cochran's private bathroom to admire the massive Lalique crystal sculpture that turned a bathing area into an art gallery.[1] Over it all, Cochran presided: directing the conversation with her large hands, yelling at cooks and staff with thunderous remonstrations, and leaving no one to wonder who was in charge. Over the next six months, Jerrie Cobb turned down five invitations to meet with Cochran, each time begging off because of urgent travel plans or unforeseen work.[2]

Although each woman realized how critical the other's support and talent were, they played a psychological tug-of-war. Cochran's muscle was exerted in trying to bring Cobb into her powerful world; Cobb's strength was displayed in persistent refusals. The well-mannered notes they exchanged belied their rising wariness.

Unaware of the friction between Cobb and Cochran, eighteen women began arriving in Albuquerque from January through August 1961 for astronaut testing. A telephone call from Dr. Lovelace's secretary or a telegram hastily sent a day or two before testing was to begin was all the advance notice most women received before they booked plane reservations. Kansas City engineer and racing pilot Sarah Gorelick was tracked down by her family while she was in a beauty shop. Her father told her that the Lovelace Foundation had called the house and wanted her for astronaut testing immediately.[3] Gorelick scrambled to get time off from work. She had already used up her annual vacation allotment for Powder Puff Derby racing and was stretching her boss's goodwill by asking for more. But Randy Lovelace had to fit the women's testing around official contractual work for the government, the airlines, or the military and could not give the women much advance warning. Most women arrived by themselves for the testing and went through the week alone.* On a few occasions, a woman reported to the foundation and discovered that she had a testing partner. Those who tested in pairs made the most of the camaraderie, sharing reactions to tests, sparking friendly competition, and often keeping in touch with each other after they returned home.[4] Candidates who came by themselves were occasionally shown around town by a Lovelace nurse during evening hours.† Otherwise, candidates were unaware of who preceded or followed them.

---

*According to records in the files of Lovelace physician Dr. Ulrich Luft, the order of testing for the eighteen women pilots who followed Jerrie Cobb was as follows: Jan Dietrich the week of January 17, 1961; Wally Funk the week of February 28, 1961; Virginia Holmes the week of March 3, 1961; Joan Merriam, Patricia Jetton, and Marion Dietrich the week of March 7, 1961; Betty Miller and Rhea Hurrle the week of March 14, 1961; Frances Bera the week of March 22, 1961; Georgiana McConnell the week of March 29, 1961; Jerri Sloan and B Steadman the week of April 4, 1961; Irene Leverton the week of April 18, 1961; Sarah Gorelick the week of June 19, 1961; Myrtle Cagle the week of June 30, 1961; Janey Hart and Gene Nora Stumbough the week of July 24, 1961; Jean Hixson the week of August 15, 1961.

†Sarah Gorelick [Ratley], who was tested alone, remembered that the female Lovelace nurses noticed that the women astronaut candidates spent all their off-hours in the Bird

Through racing circles, a few of the women heard vague descriptions of the testing whispered among friends. For the most part, however, the women who converged on the Lovelace Foundation felt oddly isolated and disconnected during the tests. Unlike during the testing of the Project Mercury astronauts or Jerrie Cobb, there were no social evenings planned with foundation staff. The women's only encounters with Lovelace physicians were during the actual testing and at the end of the week, when Randy Lovelace or a member of his medical staff reviewed the candidates' scores and told them whether they had passed or failed.

When they arrived at the shabby Bird of Paradise Motel some women must have thought they made the wrong decision. There was nothing about the motel that matched their soaring ambitions, nothing lofty or distinguished about a Route 66–type roadside motor court with linoleum floors and nondescript furnishings. The only advantage the motel offered was that it was convenient to the Lovelace Foundation, right across the street. With the dripping faucets, toilets that frequently did not flush, and countless enemas to be self-administered over the next week, inspiration was hard to find. Lucky was the woman pilot who arrived for her testing on a Wednesday. That was the one day a week that the Bird of Paradise housekeepers changed the sheets.[5]

On top of the hostility some women found from their employers when suddenly requesting time off for unexpected travel they couldn't explain, some husbands were equally uncooperative. It seemed like a pipe dream to drop child care chores and family responsibilities for something as audacious and unrealistic as astronaut tests. Ignorant of the brewing tension between Jerrie Cobb and Jackie Cochran, a few women found themselves being used as pawns in the tug-of-war for leadership of the program. Over the next eight months, more than one woman risked a job or a marriage by hurriedly stuffing a few slacks and blouses, an Agatha Christie mystery novel, and a toothbrush into a bag and leaving on the next plane for New Mexico. Passing the astronaut tests was just the beginning.

of Paradise Motel by themselves. One nurse volunteered to show Gorelick the sights of Old Albuquerque during her stay. During their tour Gorelick stopped at a jewelry store and picked out a southwestern-looking turquoise-and-coral pin as a memento of her testing. The pin looked like a bird about to take flight. Forty years later, the pin remains one of her favorite possessions. It reminds her of great dreams, she said. She calls it "an inspiration."

The Bird of Paradise must have been a particular jolt for thirty-four-year-old Jan Dietrich, who arrived for testing at the end of January 1961. She had just returned from a week at Jackie Cochran's ranch in Indio, where she had spent long days exercising and relaxing by the pool. Jan and her twin sister, Marion, knew Jackie Cochran from local flying circles and both lived not far from the Cochran-Odlum ranch.[6] Cochran took a special interest in their testing, and Jan was the first woman scheduled for exams. Strikingly attractive, the dark-haired twins bore a close resemblance to Hollywood actress Natalie Wood; some said they could be mistaken for the new First Lady, Jacqueline Kennedy. Cochran, always concerned that her WASPs conform to feminine ideals, realized the Dietrichs were especially compelling representatives of female aviators: they were both skilled pilots and glamorous-looking women.

Jan sent nightly letters back to Marion, knowing that her sister would be among the next candidates. "You'll be up at 5:30 or 6. You will be running all day long. Come with a little extra weight; you will miss one or two meals every day. . . . Because of the crash program, some of the tests will be a little incompatible. This cannot be helped. But try not to have your color portrait taken the day they rub clay all over your head for the electroencephalogram. Or take your exercise test the morning you take three enemas in two hours. I think you get the idea."[7] At the end of the week, Jan met with Dr. Secrest, who informed her that she was in the "upper 10% of the 65 astronaut applicants and test pilots who have gone through the astronaut testing program."[8] With her 8,000 hours of flying time, an airline transport rating, and a University of California at Berkeley degree like her sister's, Jan Dietrich joined Jerrie Cobb as the second woman to successfully pass the Lovelace medical tests for astronaut viability.*

Marion followed six weeks later, receiving notification of her testing date just two days before she was to report. During the previous fall, she had

*Few women in the early 1960s had earned an airline transport rating (ATR). Considered the "master's degree" of aviation credentials, an ATR requires experience with high-performance, multiengine aircraft and advanced instrumentation. Commercial airlines normally require an ATR for pilots who command their planes. In the 1960s, since virtually no women were being hired as pilots for commercial airlines, few women even sought the expensive schooling necessary to prepare for an ATR. Four other Mercury 13 women held airline transport ratings: Jan Dietrich, Irene Leverton, B Steadman, and Myrtle Cagle.

almost given up thoughts of astronaut tests since she and Jan had not heard anything directly from the Lovelace Foundation following the initial letter of invitation in September. By Thanksgiving they assumed the experiment was off. Jackie Cochran intervened and, in her new position as a consultant to the program, asked Randy Lovelace about the twins' status.* Lovelace's secretary informed Cochran that Marion and Jan Dietrich were being given "utmost" consideration and that Lovelace was particularly interested, from a scientific standpoint, in testing identical twins.[9] When Marion's call to report to Albuquerque finally came, she took time off from her job as a general reporter and feature writer for the *Oakland Tribune*. Like the astute writer she was, she paid attention to details as she checked into the Bird of Paradise and filed them away for a future magazine article. "You're really in for it," the world-weary motel manager warned her, having seen Cobb and Jan Dietrich already go through the tests. "It's rough."[10] Since getting her pilot's license with Jan as a teenager and adding extra air time with the Civil Air Patrol, Marion already had accumulated more than fifteen hundred hours and had earned seaplane and flight instructor's ratings. She devoted her weekends and vacations to flying, and ferried planes with her sister, occasionally to locations across the Atlantic. Like other serious women pilots after the war, she and Jan also participated in regional air races and even won second place in the All Women's Transcontinental Air Race the first time they entered. At five feet, three inches and 103 pounds, Marion hoped size would work to the Dietrichs' advantage. "Weight is critical," she wrote her sister. "It takes, I think, 1,000 pounds of thrust for every pound of payload," adding that their small size gave them an edge over the taller and heavier Jerrie Cobb.[11] By the end of the week in Albuquerque, Marion Dietrich learned that she, too, had passed the physical tests. What Dietrich did not fully comprehend, however, was that she and Jan were about to be used to assert Jackie Cochran's prominence in the testing program.

Even though Jackie Cochran warned Randy Lovelace to avoid focusing

---

*Pat Dietrich Daly, the Dietrich twins' sister, believes that Jackie Cochran may have initially suggested Jan and Marion to Randy Lovelace. Since Cochran and the Dietrichs were members of the California chapter of the Ninety-Nines, they knew each other fairly well, Daly believes. No other material evidence exists to indicate exactly how Jan and Marion Dietrich appeared on the testing list. Certainly once they were tapped for testing, they were favorites of Jackie Cochran. Some other members of the Mercury 13 view the Dietrich sisters as Cochran's "handpicked" candidates.

media attention on one individual candidate, Cochran did exactly that when she wrote a cover article that spring for *Parade* magazine, a Sunday supplement that appeared in major newspapers throughout the country. In the April 30, 1961, article—less than three weeks after the country focused on the startling news of Soviet Yuri Gagarin's orbit of the Earth—Cochran wrote a story about "Jan and Marion Dietrich: First Astronaut Twins" and her own involvement with the Lovelace program. The cover photograph featured the Dietrich twins standing confidently, with gleaming smiles, in the stock woman-pilot pose: in flight suits and holding borrowed Air Force jet helmets. Cochran outlined the testing being conducted in Albuquerque and illustrated the article with staged photographs of herself standing in the background with clipboard in hand watching Jan on the treadmill. While Cochran predicted that a woman would not go into space for another six or seven years, she stated that Lovelace's independent experiment "could eventuate into a government-sponsored program."[12] It was too expensive right now, she argued, to qualify women as jet pilots when no national emergency existed and when so many women allowed marriage and children to interrupt their aviation careers. Cochran proposed that qualified women pilots write her directly if they were interested in being considered for the program. Thus sidestepping Randy Lovelace and Jerrie Cobb, Cochran positioned herself as the gatekeeper for any test subjects. Marion Dietrich, who did not seem to realize that she and her sister were caught in a battle of control between Cobb and Cochran, sent Jackie a congratulatory note after the article appeared.[13] Two months later, Dietrich was bewildered and stung upon receiving a frosty response from Cochran. Marion Dietrich's notes from the Bird of Paradise resulted in an article about the testing that would soon appear in *McCall's*. Cochran had found out about the upcoming publication, did not appreciate being upstaged in print, and implied that Dietrich's publicity breach might threaten her further involvement in the program. "I understand there is some present complication in your case because of a story you have sold for publication," Cochran wrote. "I sincerely hope this problem gets straightened out."[14]

Cochran's *Parade* article generated a flood of responses from women pilots and even young girls who were inspired by the idea of women astronauts. Newlywed Myrtle Thompson Cagle from Macon, Georgia, was the only pilot whose letter to Cochran resulted in an invitation to Albuquerque for testing. Other women also sent their credentials to Cochran, but Randy Lovelace did not believe they met the basic requirements. Although the thirty-five-year-old Cagle certainly wanted to be considered for the women-in-space program, she was worried that her new husband would disapprove and waited until he left for work one morning before writing Cochran a gushing letter as one of "your most ardent fans." Cagle offered an impressive resume: airline transport and multiengine ratings, 4,300 flying hours, and work as a flight instructor at a local aero club. Randy Lovelace immediately telephoned Cagle and asked her to come to the foundation the next day. Cagle could not comply, telling Cochran that since her husband was not at home, she could not secure his permission. Cagle did travel to Albuquerque some weeks later, apparently with her husband's consent, and sent Cochran a woodpecker pepper mill from Los Alamos as an expression of gratitude for her help in submitting her name and financially underwriting her expenses. Cagle, too, passed the exams and maintained her correspondence with Cochran, whom she regarded as the leader of the program. "I want you to be the head of our group," she emphatically declared. "Already I think of you and refer to you as my 'big sister.' I don't have a sister and you are my aviation one. I don't want to sound mushy, I just admire you very much."[15]

Jackie Cochran may have wanted to play a more active role than consultant to Lovelace's women-in-space program. While in Albuquerque, candidate Sarah Gorelick found herself within earshot of a heated argument between Lovelace and Cochran just after being informed she had passed her tests. Twenty-seven-year-old Gorelick held a degree in mathematics from Denver University, with minors in physics and chemistry, and worked as an engineering assistant at AT&T in Kansas City.* As the only pilot with a technical background, Gorelick hoped to be slated for communications work if her astronaut training continued. Yet despite the good news about Gorelick's test results, Lovelace seemed distressed. Eventually he told Gorelick that he had bad news to break to the next person on his schedule. As it turned out,

---

*The "assistant" part of Gorelick's title annoyed her, since men with only high school diplomas—whom management required that she call "sir"—worked as engineers, while all the women with science degrees from universities were stuck in assistant roles. It was a relic of an unenlightened past, Gorelick acknowledged, just like the lemonade that the company poured out when the temperatures in the unair-conditioned office went above 105 degrees.

Jackie Cochran had come to the Lovelace Foundation for her annual physical the previous month. Her exam results indicated problems, and Lovelace unthinkingly muttered to Gorelick, "She's not going to be happy about this." When Cochran appeared at Lovelace's office for the next appointment, Gorelick greeted her respectfully and then quickly disappeared. As she was walking down the hallway, Gorelick heard shouting coming from Randy Lovelace's office. While only Lovelace and Cochran would ever really know what news was being delivered, Gorelick's conversation with Lovelace led her to believe that Jackie Cochran was furious that her medical reports ruled out any possibility of astronaut candidacy. At the time of the argument, Gorelick was not fully aware of the complex forces battling for control of the women's astronaut program. Like most of the women who arrived at the foundation, she was oblivious to underground power wars and focused on simply passing the exams. Intuitively, however, Gorelick sensed that she did not want to feel indebted in any way to Jackie Cochran. When the reimbursement check that Cochran had donated to cover the women's expenses while in Albuquerque arrived, Gorelick opened the letter but never cashed the check.[16]

Like Myrtle Cagle, Jerri Sloan was having trouble getting away from home. Her husband, Lou, was not at all sure he wanted his wife to go to Albuquerque. While Lou Sloan had always supported her flying (the two had gone to the University of Arkansas together for its aviation program), astronaut testing seemed too far-fetched. And Sloan had been warned that surpassing a husband's accomplishments meant trouble. "Baby, let me tell you something," her stepfather had whispered to her while they were dancing at an Amarillo country club. "The worst thing you can do is to be in competition with your husband." While her husband's disapproval seemed like envy, Sloan was beginning to understand that the real problem was his drinking. Like many of his buddies who had been World War II pilots, Lou Sloan liked an occasional drink. Over the years, however, his social drinking had become excessive, and the young man Sloan had once found loving and "too good-looking to be real" had turned remote and angry. With or without her husband's approval, Sloan knew she was going to Lovelace, and she called her mother to baby-sit the children for a week.[17]

When Jerri Sloan met tall, slim Bernice "B" Trimble Steadman at the Bird of Paradise, she was grateful to have a testing companion. She was even glad for the tests themselves, a welcome diversion from the problems at home. Sloan had never met Steadman even though both women were active in the Ninety-Nines and in women's air-race circles. Steadman lived with her lawyer husband in Flint, Michigan. Thirty-five years old, she had worked her way up from an AC spark plug inspector to a professional pilot with eight thousand hours in the air and owner of Trimble Aviation, a fixed-base operation providing general aviation servicing instruction, repair, and sales. Calm, patient, flexible, considerate, Steadman always seemed to take the long view. When she was a year and a half old, a house fire had swept though the Trimbles' home in Michigan's Upper Peninsula. Although she and her parents had gotten out safely, her father had gone back into the burning house to rescue two other children. He did not return and perished along with her two older siblings. Steadman was too young to remember details of the tragedy, but the experience affected her family profoundly and marked her life as well. Perhaps it was because Steadman had experienced such a traumatic beginning that every other hurdle in life seemed small by comparison.[18]

B Steadman's steadying influence helped Jerri Sloan focus on passing the tests, especially when calls started coming in to the motel from an angry Lou Sloan. Often he would call drunk, ranting incoherently until Sloan would hang up. Steadman knew Lou Sloan's calls were making testing difficult for his wife. The fact that Steadman's husband, Bob, was proud of his wife and even took on the chore of moving the family by himself from Flint to Montrose, Michigan, during the testing offered a contrast to Lou Sloan's unsupportive behavior. Steadman wondered how Sloan could concentrate on the tests with her husband's phone calls amounting to nightly harassment. Randy Lovelace also knew about Sloan's strained marriage and the pressure she was receiving from her husband to drop out. Lou Sloan called Lovelace's office as well, trying to derail his wife's testing or at least give Lovelace reason to question her further involvement.[19]

On their final day at the foundation, both Steadman and Sloan received the news that they had passed the exams. Lovelace called Sloan into his office for an additional private conversation. He asked her if problems in her marriage would prevent her from taking the next step in testing. Sloan emphatically told Lovelace that any difficulties she was having had been simmering long before she received the call from Albuquerque. As much as she loved her husband and as much as she believed he still loved her, his drinking had made their life miserable and she knew what she had to do. She told Dr. Lovelace that she wanted to be included in the next phase of testing and would take care of personal matters at home. When she stepped off the plane back in Dallas, Sloan was not prepared for what awaited her. A lawyer's courier met her at the gate and presented her with divorce papers. At home she found her husband in bed and drunk. Her dreams of spaceflight seemed very far away as she packed up her children and her mother and drove off to find a motel for the night.[20]

Even though Wally Funk had not been on the original list of candidates and had submitted her own name, Randy Lovelace believed she merited consideration. Like the Dietrich sisters, who intrigued physicians because they were identical twins, Funk sparked Lovelace's medical curiosity because at twenty-two years old she was the youngest of the invited group and she was remarkably fit from competitive skiing. Plus, she had been raised in altitudes above 8,000 feet in Taos—a possible physical advantage for high-altitude spaceflight. Doctors in Albuquerque looked upon Funk's candidacy with a measure of local pride as well. She was the only person to undergo astronaut testing at the Lovelace Foundation who was a native of New Mexico. Determined to beat the scores of all the other women but aware that her youth and lack of sophistication might work against her, Funk kept to herself during her testing and asked few questions. Only when she was truly baffled would she let herself reveal her inexperience. That was the case when a clinician asked Funk for a stool sample; Funk had never heard the term and had no idea what to do with the small plastic cup.[21]

Wally Funk's naïveté, combined with her fierce determination to rack up the top score on every test, paved the way for a serious misunderstanding, a false impression that remained unexamined for years. The confusion centered on Funk's performance on the bicycle test and her later public assertion that she had beaten John Glenn, surpassing the celebrated astronaut in fitness. The goal of the test was to keep pedaling a stationary bike that felt as though it were going up steeper and steeper hills. Funk's blood pressure, respiratory volume, and respiratory gas exchange were measured each minute, as her heart rate approached a maximum of 180 beats a minute—the point of total physical exhaustion. Determined to surpass the maximum time that anyone had stayed on the bike, Funk asked Lovelace technicians for the top time: ten minutes, she was told. As she pushed toward ten, Funk felt her legs grow weak and her heart race. She dug in, found a second wind, and pushed to eleven minutes before refusing help in getting off the bike and crumpling to the floor.

What Funk did not realize, however, was that her total time spent on the bike was only part of the equation to determine fitness. Her time would be calibrated with her age, weight, and other metabolic factors to calculate overall aerobic capacity. Duration on the bike, while important, was not the sole indicator of success. While both John Glenn and Wally Funk delivered an exceptional performance on the bike test, Funk did not beat Glenn and in fact did not hold the top female score.* She came in third among the women. Jan Dietrich had the highest score among the women, followed by Rhea Hurrle.† While her eleven minutes on the bike certainly showed her determination, Funk's misunderstanding also suggested a tendency to overlook details, especially if they contradicted the image of a top competitor that she clearly wanted to project.

For Wally Funk, Jerri Sloan, B Steadman, Sarah Gorelick, and all the other women pilots who were undergoing testing at the Lovelace Foundation, there was no question that they believed the examinations could lead to becoming an astronaut. They studied Dr. Lovelace's invitation letter with its reference to "the women in space program" and reasoned that any man whom NASA tapped to help select the Project Mercury astronauts certainly would be taken seriously. Besides, Jackie Cochran, in her *Parade*

---

*Dr. Jack Loeppky inherited the medical files of his mentor, Dr. Ulrich Luft, the Lovelace Foundation pulmonary specialist who administered the bicycle test to male and female astronaut candidates. Luft's notes indicate that Glenn spent 17 minutes on the bike and his peak oxygen consumption was 2.801 liters per minute. His weight was 83.5 kilograms. In comparison, Funk spent 11 minutes on the bike and her peak oxygen consumption was 1.812 liters per minute. Her weight was 57.4 kilograms. When the scores were computed for maximum oxygen used by body weight, Glenn's score was 33.5 milliliters per kilogram and Funk's score was 31.6. Loeppky interpreted the scores as indicating "precisely—he [Glenn] was a bit fitter." Loeppky's files from Dr. Luft are the only records that have survived that compare the scores of all nineteen women (including Jerrie Cobb) who were tested at the Lovelace Foundation. Lovelace Foundation and Clinic historian Jake Spidle, Jr., believes that many of the medical records associated with the nineteen women who underwent astronaut testing in 1960 and 1961 were lost, destroyed, or "in a landfill somewhere."

†The rank order on the bicycle test of all nineteen women who underwent testing at the Lovelace Foundation is as follows, starting with the top score: Jan Dietrich, Rhea Hurrle, Wally Funk, Jerrie Cobb, Marion Dietrich, Jerri Sloan, Janey Hart, Myrtle Cagle, Gene Nora Stumbough, Jean Hixson, Virginia Holmes, Joan Merriam, Betty Miller, Bernice Steadman, Frances Bera, Patricia Jetton, Georgiana McConnell, Irene Leverton, and Sarah Gorelick.

magazine article, said that the women who passed the Lovelace tests "may later receive specialized training to participate in space flight as astronauts."[22] Only one woman among the Mercury 13 questioned that the Lovelace tests might lead to spaceflight. Gene Nora Stumbough doubted any of the women would become astronauts, and believed that the tests merely were part of a scientific research program. Her perspective stood in contrast to Lovelace physicians such as Dr. Donald Kilgore, who was convinced that the top candidates among the women would go on to become astronauts. Forty years after her testing, Stumbough argued that she did not have the qualifications to join the space program. "I was a minimal-hour pilot," she said. "We had no qualifications to be astronauts. No scientific degrees. We hadn't been test pilots or done advanced training." Yet when she checked into the Bird of Paradise Motel for the week's testing, Stumbough later remembered feeling the pressure of performing well. "You were carrying everybody else on your shoulders," she said. "You felt like you had to make a good showing for the ones who followed."[23]

Jane Briggs Hart arrived at Lovelace the same week as Stumbough and did not remember her testing partner's skepticism about the program. Why would anyone go through with the week's difficult testing, Hart later asked, if she did not think it might someday lead to spaceflight? At forty years old, with an active life in Washington, D.C., supporting her husband's liberal causes in the U.S. Senate, and with eight children at home between the ages of four and fourteen, Janey Hart was not interested in joining Lovelace's project if it only meant being a lab rat. She was interested in the space program and thought her nineteen years of flying, helicopter pilot's license, and service as a captain in the Civil Air Patrol made her a credible astronaut candidate. With the support of her husband and children, Hart prepared for the week in Albuquerque with carefully orchestrated plans for her absence. She loaded up the freezer with roasts and vegetables and arranged for milk delivery three times in the upcoming week. She pulled three carts into the grocery store checkout line and warned people behind her to choose another line, since she would be there for a while.[24]

Dr. Donald Kilgore, who administered her ear tests at the Lovelace Foundation, was immediately struck by Janey Hart's self-possession and no-nonsense attitude. He was also impressed by her motivation, recognizing that there were formidable forces and prejudices to overcome for any woman who wanted to fly. More than one woman who flew charter work, for example, reported that passengers would leave the plane when they realized the pilot was a woman. Stumbough's mother was called "unnatural" for allowing her daughter to fly. Janey Hart's devotion to her children was questioned when word got out that she wanted to be an astronaut. For single women, bias often took the form of insinuations that they were lesbians because they were working in a "man's field." The unspoken edict that women pilots in the 1950s fly in high heels and be coiffed and well dressed when they landed was embedded in the assumption that women who wore makeup and stockings were "ladies" and therefore "normal." After sending along a photograph with her application for a job as a demonstration pilot for Beech Aircraft, Stumbough was contacted by a prospective boss who told her, "If you are as feminine as the picture appears, we'll hire you."* Women who were indeed lesbians were forced deeper into the closet—often for a lifetime—terrified that they would lose their jobs, friends, and the camaraderie of the airport hangar, which often substituted as family.[25]

Not every woman passed the Lovelace tests. One candidate feared that claustrophobia in the Los Alamos radioactivity counter, as well as comments she made about family and business obligations, had ruled her out.[26] A memo forwarded from Floyd Odlum to Jackie Cochran told another story: "The woman who failed (heavy smoker—heart did not stand up to bicycle test, etc.) will be given another chance if she cuts down smoking and exercises."[27] The candidate later received a letter of appreciation from the Lovelace Foundation but with no invitation to further testing.[28] Two other candidates had previously undiagnosed brain abnormalities detected while at the foundation, though neither woman ever developed neurological problems.[29] Another woman came down with sinus problems during the exams and was not invited to continue.[30] Two other women never knew the exact reasons why they did not pass the exams, learning only in a letter from Randy Lovelace after they returned home that they did not meet the requirements of the women-in-space program.†

---

*Gene Nora Stumbough Jessen recalled that the job at Beech Aircraft that she had accepted in 1962 turned out to be her "dream job." Few women were being hired as demonstration pilots in the early 1960s and she considered herself fortunate to serve as a member of Beech's "Three Musketeer" demonstration team flying all over the country.
†Virginia Holmes was the candidate who believed that claustrophobia had ruled her out of consideration. Fran Bera, a pilot with more Powder Puff Derby wins than any other

The secrecy about the tests that Randy Lovelace insisted upon also took its toll. One candidate, thirty-year-old Rhea Hurrle, never told a single member of her family about her astronaut testing. She had never even mentioned to them that she was an exceptional pilot who participated in air races and had a budding interest in seaplane flying. A secretary and executive pilot for a small aircraft sales and engineering firm in Houston, Hurrle kept her aspirations to herself. Two years later, her parents unexpectedly found out she had been tapped for astronaut tests when *Life* magazine featured their daughter's photograph along with the other women's in a multipage spread. "I was going to wait until I got into space" to tell them, she later said.[31]

When she received the call to come for astronaut tests, Irene Leverton was working at a Santa Monica fixed-base operation, flying charter and offering flight instruction. Her boss refused to give her the week off, and Leverton asked Lovelace for an alternate date. Lovelace reworked the foundation schedule and proposed that Leverton come several weeks later. Deciding the opportunity to become an astronaut was worth the employment risk, Leverton told her boss's wife that she would be away for a week and left for Albuquerque. When she returned, having passed the exams with

woman, was told by Lovelace physicians who examined her that they had discovered "abnormal brain waves." Concerned that she had a serious health problem, Bera had two subsequent evaluations once she returned to California but never did encounter neurological problems. Pat Jetton and Joan Merriam were tested at the same time. Like Bera, Jetton was told that test results indicated a brain abnormality, although doctors assured her it was nothing to be concerned about and that it might never manifest itself during normal daily activities. Nevertheless, when Jetton returned home to Dallas, she consulted a neurologist, whose tests, while not, Jetton admitted, as extensive as those administered at Lovelace, revealed nothing. Merriam, twenty-four years old, was a pilot for a construction firm and the youngest woman in the country with an airline transport rating. She had more than six thousand hours of flying time. Merriam was informed by Lovelace physicians that she had not passed the tests. She later was killed in a crash of her airplane. Betty J. Miller had sinus problems as she went through the tests, and when Lovelace physicians suggested they perform surgery on her nose right then, Miller opted to consider the matter further. Georgiana McConnell thought she had done well on all the tests with the exception of swallowing three feet of rubber hose for gastric analysis. She knew she had a strong gagging reflex and had difficulty getting the hose down her throat. When she returned home to Nashville, McConnell privately recorded her thoughts about the previous week's testing. "I would never have forgiven myself if I had turned down an opportunity to see if I were capable of being one of these chosen few," she wrote. Aside from receiving a letter from Randy Lovelace thanking her for her participation but stating that she did not meet astronaut qualifications, McConnell never learned any more details about why she was not invited to continue.

a caveat from Lovelace to drop some weight, she found that her job had changed. The charter work she enjoyed had suddenly dried up, and she was stuck giving flying lessons to beginning students in slow planes. For the thirty-four-year-old pilot with nine thousand flying hours behind her, a nearly completed airline transport rating, and experience as a gutsy forest service pilot fighting fires over the Sierra Nevada, it was a huge demotion. Angry and downcast, Leverton moved to Los Angeles, where fellow Ninety-Niner Jan Dietrich tried to help Leverton find work and even opened her home to her for a few weeks when Leverton found herself virtually on the streets. Explaining that she had passed the same physical exams as the Project Mercury astronauts might have opened doors for Leverton in Los Angeles's competitive aviation field, but she kept her testing confidential as she had promised Dr. Lovelace. Eventually Leverton found a job but never mentioned Albuquerque even when her company hired a young woman with an unusual name: Wally Funk.*[32]

AS THE SUMMER OF 1961 PROGRESSED, NEWLY APPOINTED NASA chief James Webb made the round of Sunday television interview programs and was asked if he was aware of female "recruits" currently undergoing testing in Albuquerque. Webb admitted he was briefed on the program and mentioned Jerrie Cobb, specifically, as having set a light plane altitude record the previous year and her interest in becoming an astronaut. There were as many as a dozen more taking part in training at Dr. Randy Lovelace's private clinic—the same clinic, Webb pointed out, that had put Alan Shepard and the other Project Mercury astronauts through rigorous physical tests. "We haven't got that far in our planning," Webb admitted to reporters who asked him if the United States would be launching a woman astronaut soon.[33]

Randy Lovelace, however, already was preparing for the next step. As intent as he was to gather more scientific information on the pilots, the

---

*The two women differ on the extent to which they acknowledged their mutual experience at the Lovelace Foundation. Leverton remembers that she never discussed the testing with Wally Funk while they were both employed at Hawthorne Aviation in Los Angeles. Funk recalls that at Hawthorne she did eventually discover that Leverton also had been tested, although they barely discussed the topic, both concerned that others might find out and publicity could damage the program.

Lovelace tests had determined that women could perform as well as men on examinations to determine endurance, resilience, and physical capability.* As many Lovelace Foundation physicians later observed, the tests of women pilots in 1960 and 1961 were among the first comprehensive medical exams ever performed on *healthy* women.[34] Most previous medical tests had focused on either men or women who were stricken with a specific disease to be studied. The Lovelace tests, however, took the unprecedented stance of evaluating physically fit women. The initial results demonstrated that these women were not fragile, weak, or prone to physical vulnerability when compared with men. Lovelace determined that women had no inherent, biological, or physical limitation that would prevent them from operating as well as men in the extreme conditions of spaceflight. That conclusion flew in the face of prevailing societal assumptions that women were the weaker sex. Although the test scores did not immediately turn around society's deeply intractable myth of women's physical inferiority, they served as a critical early challenge to the prevailing wisdom of the scientific community, which claimed woman was a lesser form of man.

Just as important as the conclusions on women's physical ability, the Lovelace tests also demonstrated that women have a desire to explore the unknown. Confronting risk, calculating danger, even seeking adverse situations that tested one's confidence and skill were not characteristics of men alone. Irene Leverton was familiar with men who wanted to exclude her from challenging flight assignments. "Some men think," Leverton said, "that by not allowing me in, they become the only ones with courage."[35] Just as the testing exploded the myth of women's physical limitations, so too did it undercut the idea that all women were content with a placid or confined

---

*The selection process for men and women who underwent astronaut testing at the Lovelace Foundation provides an interesting contrast. Jerrie Cobb reviewed the credentials of eight hundred women commercial pilots. Randy Lovelace invited twenty-five to be tested. Nineteen underwent testing, and thirteen women passed. A total of 508 U.S. Air Force, Navy, and Marine personnel were initially screened for astronaut potential, and 110 met minimum standards in terms of test pilot training, flight hours, age, height, weight, and technical education. Sixty-nine men reported for initial briefings in Washington, and after a second round of reviews, thirty-two men were sent to the Lovelace Foundation and Wright Aeromedical Laboratory for medical and spaceflight simulation tests. After exams, eighteen men were recommended to NASA without medical reservations. Eventually seven men were selected as Project Mercury astronauts when the NASA selection committee could not agree upon six final candidates.

life. The thirteen women who successfully passed the Lovelace tests wanted to demonstrate their mental and physical capabilities. They wanted to be challenged as pilots. They wanted to contribute their talents and even their lives to a national effort. By the end of August, Randy Lovelace could confirm that thirteen American women had passed the same physical tests given the Project Mercury astronauts. While he promised the Mercury 13 nothing, Lovelace was as eager as they were to take the next step.

CHAPTER SEVEN

Project Venus

AFTER EACH OF THE MERCURY 13 RETURNED HOME, RANDY LOVELACE
sent an official letter of congratulations and asked if she would be willing to
proceed to the next phase of testing. Everyone answered yes. Lovelace still
was unclear exactly where that new testing would take place, although he
knew he wanted to test the women as a group at a military laboratory where
they would experience spaceflight simulation tests in a centrifuge and an
altitude chamber, perhaps even spend some time piloting jets. Collaborating
with Wright-Patterson Air Force Base was out of the question, even though
Lovelace had extensive contacts at the facility from his time as chief of the
Aeromedical Laboratory. The resistance Don Flickinger had experienced
two years earlier when he had wanted Jerrie Cobb to start her astronaut
testing at Wright made clear the Air Force's position on women astronaut
candidates. What Flickinger termed the "unfortunate Nichols release" had
destroyed any chance for cooperation. As Lovelace worked to secure a testing
site, he urged the women to work on their conditioning since he predicted the
upcoming tests would require considerable physical stamina.[1]

Meanwhile, Jerrie Cobb had ideas of her own. Cobb knew that after
the Project Mercury astronauts had finished with physical testing in
Albuquerque, they proceeded to Wright-Patterson for two additional phases
of testing: spaceflight simulation exercises and a comprehensive psychiatric
exam to determine mental fitness for spaceflight. With Lovelace negotiating
a site for simulation testing, Cobb found a laboratory for psychiatric evalua-

tions. Just blocks from her home in Oklahoma City was an excellent alternative to the Wright-Patterson labs. At the Oklahoma City Veterans Hospital, eminent psychiatrist Dr. Jay Talmadge Shurley served as founder and director of the Behavioral Science Laboratory; he was also professor of psychiatry and behavioral science at the University of Oklahoma College of Medicine. Shurley's innovative experiments were well known among residents of Oklahoma City, and his prestige among scientists was nationally acknowledged. A tall, affable Texan, Shurley was familiar with Cobb's aviation accomplishments, and the prospect of assisting with the women-in-space program was enticing. After he received an inquiry from Cobb, Shurley consulted with Randy Lovelace and agreed to start "phase two testing," a term NASA used to describe psychiatric evaluations of astronauts.

By his own admission, Shurley was the "enfant terrible" of his profession.[2] At a time when psychiatry was still regarded in some medical circles as an illegitimate science, Shurley pushed original thinking even further. The University of Oklahoma was the first in the country to start an academic program in behavioral sciences—merging the study of medicine, the science of the mind, and analysis of behavioral forces that affect mental health. "It was the relationship of the mind and the body at all times," Shurley said regarding the focus of his study. At the beginning of his career, Shurley served as the first chief of adult psychiatry at the National Institute of Mental Health (NIMH) in Bethesda, Maryland. There he met scientist Dr. John C. Lilly, who was conducting groundbreaking sensory isolation experiments as chief of the Cortical Integration Department of the Neurophysiological Laboratory. In a silent, odorless, dark room, Lilly had constructed a large water tank—"a Lilly pond," Shurley called it—in which subjects floated alone until they reached the limit of their tolerance. The tests shed light on how the lack of stimulation affected the human mind and the ways in which hallucinations were produced naturally when the mind was not stimulated. Shurley expanded on the tests, applying them to pioneering work on schizophrenia, discovering that schizophrenic patients ceased having hallucinations when they were placed in the tank. Shurley surmised that when a schizophrenic's external world was removed and all stimuli were absent, the patient no longer needed to produce hallucinations. Producing hallucinations was a schizophrenic's defensive mechanism to keep the external world at a distance.[3] After some members of Congress learned that Lilly and Shurley were conducting experiments that caused mentally healthy people to hallucinate, the NIMH ordered the physicians to halt their study. Although their superiors at NIMH acknowledged that Lilly and Shurley were conducting valuable science, they also realized they could not alienate politicians and their source of funding. The work was too innovative for its time. Eventually Lilly moved to the U.S. Virgin Islands, where he began inventive work studying communication between humans and dolphins. Shurley accepted the job in Oklahoma and a prestigious position as senior medical investigator with the VA. The position, one of five in the country, was the first the VA ever awarded to a psychiatrist, and it provided Shurley free rein when it came to deciding what to study. "It was like Christmas 365 days a year," he said.[4]

When Cobb approached him with the idea of administering psychological and psychiatric tests for astronaut viability, Shurley's curiosity was piqued. In a speech he had delivered two years earlier on the psychology and physiology of spaceflight, Shurley said he wanted to test how sensory isolation could be applied to the assumed silent void of space. He also said he had a hunch the isolation tank could be used to simulate weightlessness. Shurley had been a consultant on the isolation experiments conducted on the Project Mercury men at Wright-Patterson and found those studies not comparable to the greater isolation produced by the tank he and Lilly pioneered. Shurley estimated that fifteen minutes in his tank was the equivalent of two to three *days* in the silent room at Wright. What Shurley termed Wright's "dry-air studies" of isolation were simply not as advanced as those conducted in Oklahoma City. When NASA sent him a fact sheet on the psychiatric testing procedures for the Mercury astronauts, Shurley found the descriptions vague and lacking in any evaluation of simulated weightlessness. He circled the summary of the isolation experiment and penned above it *"NOT* a meaningful test!"[5]

Cobb's request to undergo psychiatric tests appealed to Dr. Shurley on many levels. He knew Randy Lovelace and Don Flickinger through medical circles and was familiar with their reputations as innovators and men who were intrigued by the emerging medical field of behavioral studies. "Randy reminded me of me," Shurley later said. Moreover, Lovelace was one of the few people in authority in the space program who gave any thought to the possible effects of too much and too little stimulation on the human mind.[6] Lovelace's interest in women as medical subjects caught the psychiatrist's attention as well. Shurley had long made a point of using mentally healthy women in his experiments—a practice just as novel in psychiatric circles as it was in other fields of medicine. He had initially raised eyebrows by doing comparative studies of male and female behavior during sensory isolation, but the Veterans Administration "had gotten over the shock," he said, of his unique research and superiors did not forbid him from setting out on uncommon paths.[7] As innovative as Shurley's previous experiments had been, few would be more pioneering than the one upon which he was about to embark with the Mercury 13. In evaluating Cobb's willingness to place herself in the potentially dangerous environment of outer space, in calculating her motivation, her response to risk, her ability to adapt to changing circumstances, to overcome anxiety and face fear, Jay Shurley was raising an unprecedented question. Randy Lovelace had determined that women had the physical capacity for spaceflight; Jay Shurley would find out if they had the courage.

Cobb's testing took three days and was administered by Dr. Shurley and his assistant Cathryn Walters, an OU graduate student in psychology who was studying differences in the ways in which men and women responded to the stress of underwater sensory deprivation experiments.* The interest in sex differences that both Shurley and his assistant shared grew out of their recognition that medical science rarely studied women's health and had collected virtually no data on comparative studies of men and women. "There was blindness to the fact that women and men were different," Shurley said, "but that difference didn't make women inferior."[8]

The psychological tests Cobb took over the first two days represented the standard repertoire of 1960s psychological scrutiny: general intelligence exams, occupation preference analyses, and the Minnesota Multiphasic Personality Inventory (MMPI), a set of 561 personal questions designed to determine if an individual had significant character or psychological problems. The MMPI was the most used psychological test at the time, but according to Shurley it had limitations. It was "like using a fishnet with very big

---

*In a paper delivered at a 1961 VA Medical Research Conference in Cincinnati, Walters found that males "tended to be less introspective and more stimulus-oriented than females . . . and thus appeared more field-dependent than females." Walters defined "high field dependence" as a tendency "to lack insight, to repress their impulses, to be passive, to yield to their inferiority feelings, to be tense and to show low bodily evaluations, whereas persons of low field dependence tend to show self awareness, to be active, to deal with inferiority feelings in a compensatory way and to have high bodily evaluation."

holes rather than a fine skein," he said.[9] Cobb also took the classic Rorschach test, underwent an EEG and neurological examinations, and responded to a psychiatric direct examination and an informal interview with Dr. Shurley that covered her childhood, adolescence, and occupational attitudes.* The psychological evaluation of Project Mercury astronauts covered much of the same ground and included a total of thirty hours of interviews, psychological tests, and stress experiments.[†] Cobb's three days with Dr. Shurley amounted to slightly more time.

On her final day of testing, Cobb took part in an experiment more psychologically challenging than any test the Project Mercury men encountered. For the men's isolation test at Wright-Patterson—the test that Shurley found so inadequate—they were confined to a silent and dark room for two or three hours. Wright physicians admitted the scientific limitations of the test, but they also asserted that the duration and the environment of the test were sufficient for "identifying subjects who cannot tolerate enforced inactivity, enclosures in small spaces, or absence of external stimuli."[10] Fifteen of the prospective male astronauts who experienced isolation in the room later said they "programmed" their thinking, concentrating on specific mental tasks such as counting or memory games. A few men grew restless, impatient, and demonstrated a need for a structured use of time. Sixteen men allowed random thoughts to enter their minds. Most men slept, at least for a while, and believed that appearing alert and calm was the best way to pass the exam.[11] John Glenn's experience in the Wright-Patterson isolation chamber was typical. By feeling his way around the darkened chamber, Glenn located a desk and then discovered a writing tablet left in the desk

*The complete list of Cobb's phase two psychological and psychiatric testing included the following specific examinations: Wechsler, Kuder Personal Preference, Who Am I?, IPAT Anxiety Scale, MMPI, Figure Drawing, Sentence Completion, Rorschach, TAT, and Personal Perception Assessment.

†Psychological tests administered to Project Mercury men included: Rorschach, TAT, Draw-a-Person, Sentence Completion, MMPI, Who Am I?, Gordon Personal Profile, Edwards Personal Preference Profile, Shipley Personal Inventory, Outer-Inner Preferences, Pensacola Z-Scale (a test of authoritarian attitudes), Officer Effectiveness Inventory (desirable personality characteristics found in Air Force officers), and Peer Ratings. Project Mercury men also took tests to evaluate specific aptitudes, including Raven Progressive Matrices, Doppelt Mathematical Reasoning Test, Engineering Analogies, Mechanical Comprehension, Air Force Officer Qualification Test, Aviation Qualification Test, Space Memory Test, Spatial Orientation, Gottschaldt Hidden Figures, and Guilford-Zimmerman Spatial Visualization Test.

drawer. Sitting down on a chair with a pencil that he had tucked into his shirt pocket, Glenn scribbled eighteen pages, making random lists about exercise and clothing, tracing one line to the next by sliding his finger along the paper. What Glenn found engaged his mind the most was composing poetry, and he wrote seven stanzas. The final verse read:

> *Then use all your inborn talents,*
> *Use them each and every day.*
> *Add to mankind's store of knowledge,*
> *Make them glad you passed this way.*[12]

When the lights came up three hours later and Glenn emerged from the chamber, Wright-Patterson physicians believed they had appropriately measured his ability to adjust to the emptiness and silence of space.

By contrast, Cobb was about to face the sensory isolation tank. Other research subjects who submitted to the test before Cobb reported that floating in Shurley's tank was both boring and transcendent. One man described the experience as soothing, comfortable, monotonous "yet with an air of fascinating mystery." Subjects imagined coins in the bottom of the tank, brilliant golden mushrooms, terra-cotta-colored sculptures resembling Mayan or Aztec carved stone heads. One had the sensation of being a spoon in a giant glass of iced tea, being swirled around in loopy circles. Another smelled garlic. One smelled hot road tar. One heard shorebirds crying. For others the images conjured up were startling, even emotional at times. Many reflected on past experiences, calling up memories that were difficult to shed once the test was over. One subject recalled an apricot tree he used to climb when he was eight years old. Another heard his father call his name.[13] Dr. Shurley had himself spent hours in the tank in order to fully acquaint himself with the possibilities his subjects might face, feel, hear, or taste. Shurley discovered he could withstand several hours in the tank. "I found that if I spent more than four hours in there, and I have, I would be so preoccupied with what my mind dredged up, I wouldn't be worth much the rest of the day."[14] When a local television station in Oklahoma City ran a documentary on Shurley's tank experiments, the station was flooded with telephone calls. Many viewers found the experiment deeply troubling, even terrifying. It disturbed them to learn that scientists were delving so deeply into the human mind. A test subject who thought he saw coins at the bottom of the tank was one thing, but what if a person actually lost his mind?[15]

IN EVALUATING COBB'S POTENTIAL AS AN ASTRONAUT, SHURLEY HAD several specific goals in mind for her time in the tank. He wanted to evaluate her reaction to and tolerance for extreme monotony and solitude, her levels of anxiety or conflict in isolation, and her patterns of psychological defense. He also wanted to see if Cobb could refrain from action without engaging in motor activity as a way to minimize any anxiety. In addition, Shurley wanted to observe her tendency to somatization of anxiety: manifesting physical symptoms as a response to stress. Cobb's reactions to all these potential issues would indicate her likely response to space solitude, weightlessness, and diminished G forces. Using water immersion for simulating outer space was not a new idea. As early as the turn of the century a Russian scientist had suggested it. Surprisingly, however, no one actually put the idea to use until Shurley and Lilly.[16] The actual tank Cobb would use was only the second one in the world to be constructed and was built by an Oklahoma oil field welder in the late 1950s. It was modeled on the original tank John Lilly had created at the National Institute of Mental Health that was no longer used after both Lilly and Shurley departed. The new tank was housed in a two-room laboratory in the basement of the Oklahoma City Veterans Hospital, one floor below the dog kennel laboratory. The exterior room housed reel-to-reel tape recorders, audio speakers from the tank, flashlights, dials to regulate the tank's water temperature, and an aluminum lawn chair on which either Jay Shurley or Cathy Walters sat and jotted notes. No window existed between the observer's room and the tank. The only means by which a subject could communicate with the observer was by talking: a sensitive microphone was suspended above the tank, which picked up sounds as quiet as measured breathing, sighs, or whistling in the dark.

Shurley's assistant described the tank room as "a bomb shelter," with thick insulated walls and a single heavy door leading into the observer's station.[17] In the center of the room was the large circular tank, ten feet in diameter and eight and a half feet deep. Inside was slowly rippling water set at 93.5 degrees Fahrenheit—a temperature precisely chosen so that the subject was not able to distinguish between his or her own body temperature and the water itself. It was as if the body and the fluid environment became one continuous medium. Shurley found the temperature, along with the buoyancy created by Epsom salts, created a kind of Great Salt Lake—what he called a "uniform tactile field"—that simulated weightlessness.[18] The flowing water also allowed for the continuous removal of bodily wastes. Subjects were instructed to relieve themselves so that any physical discomfort related to bodily needs would be eliminated.

The tank room was insulated for sound so that any noise was silenced. Barking dogs, activity down the hall, noisy pipes—even the distant rumble of trucks outside the hospital was quelled. The room was so silent that subjects reported they could hear the sound of their own hearts beating or peristalsis of the intestines. Occasionally one would hear the eerie slide of tendons being stretched. One physician reported hearing the snapping sound of his own heart valves closing at the end of each rhythmic beat.[19] The human body, Shurley observed, "is a real noise machine."[20] Other precautions were taken to eliminate light, odor, pressure, vibration, heat, cold, and any sources of gravity in the tank. A failing of the Wright-Patterson isolation room was that the Project Mercury men never had the sensation of free-floating suspension: they could stand, walk, even sit in a chair just as they did anywhere else on Earth.[21] In the early stages of the Oklahoma City tank runs, the subjects also wore a mask pulled down over the entire head so that they could float facedown in the water. The mask allowed for inhalation and expiration of air, but it often leaked and small pools of water would creep into a subject's ears and disrupt her solitude. For Jerrie Cobb's run, Shurley used foam flotation pillows behind her head and hips, which allowed her to drift head-up. Forgoing the mask might well have added to Cobb's peace of mind. It was a grotesque rubbery hood that resembled those worn by prisoners headed to execution.

Cobb and other subjects took a tour of the tank room with all the lights on before undergoing the test. Cobb could peer into the tub and even check for coins at the bottom of the tank if she wished. She was instructed to speak during her hours in the tank if she wished to, to say out loud any thought or musing that entered her mind, and to indicate at any time if she wanted to get out of the water. Cathy Walters would sit in the lawn chair during Cobb's run because Dr. Shurley determined that Cobb seemed more at ease with a woman in a therapeutic situation. Walters would elect whether or not to answer any comments Cobb chose to make while in the tank. Walters did not want to engage in conversation during the experiment, but she or Dr. Shurley would be immediately responsive to any urgent needs. After entering the tank room in her bathing suit, Cobb would be instructed to remove her suit so that the floating experience would be as natural and elemental as possible. Cobb would stay in the tank for as long as she chose, and after she called an end to the test, she would emerge, have her vital signs taken, and complete additional tests to determine her degree of spatial disorientation, response to weightlessness, and any changes in her perception of color, shape, or texture. She also would be asked to estimate how long she had been in the tank. Many subjects who had been floating for perhaps six hours thought that their time in the tank was only about three hours.[22]

Apart from its usefulness in simulating the environment of spaceflight, the tank also offered insights into larger, more philosophical questions regarding how men and women relate to their environment—both the environment they touch and see and the interior environments of their own minds. In a 1961 landmark paper given at the Third World Congress of Psychiatry, Shurley observed that experiments in the tank illuminated the vast reaches of the human mind and underscored Sigmund Freud's assertion that "It is the much abused privilege of (perceptual) conscious activity, wherever it plays a part, to conceal every other activity from our eyes."[23] In many ways, we have it backward when it comes to hallucinations and reality, Shurley contended. "In reality, our brains are constructed to produce hallucinatory material all the time," he argued. We are constantly hallucinating deeply in our brains. The effects of sight, hearing, and other external input recalibrate the natural hallucinations our brains create. The eyes and ears produce concreteness in a very nonconcrete world. Shurley believed that the range of phenomena accessible by the human mind was far greater than what society allowed and accepted. When given freedom and even relief from the billions of daily stimuli that bombard the human mind, Shurley believed that men and women discover "sources of new information from within."[24] Floating in the tank, some people found this new information pleasant, even exhilarating. Others confronted the unexpected images with fear and denial.

Just before Cobb was set to take her run in the tank, a reporter contacted Dr. Shurley and asked if he might try the experiment in order to provide the public with a firsthand account. Just like military officers in peacetime who considered it good public relations to allow civilians to experience jets or space simulation tests, Shurley agreed to let the reporter take a run. The journalist's record provided an unusually detailed account of the isolation tank's effects on the mind of a mentally healthy person. It also offered a striking contrast to what Cobb would soon experience and what she would say or not say into the microphone hanging from a thin cord above the dark waters of the tank.

After hearing preliminary information about what to expect in the inner room, the twenty-nine-year-old reporter spent his first half hour motionless and engaged in an audible monologue about work that needed to be completed on the job, his son's well-being, and his wife's recent sleepwalking episode. He then turned his thoughts to an unexpected letter from a former girlfriend, criticisms concerning the younger generation of journalists, and childhood memories. By the second hour, he spoke of the need to move and to exercise and expressed surprise that he did not want a cigarette. He told of his stark loneliness except for "my very real companions, my thoughts and memories." He also expressed empathy for Sam, the NASA monkey who recently had been launched into space and endured what the reporter considered to be a similar situation. The reporter then passed the time by whistling and dropped off into a brief nap in which he remembered the image of "a sawdust cream cone." By the third hour he thought he heard dogs barking and a crackling sound, and he launched into a bawdy rendition of a barroom song. For a moment he was in an ecstatic mood and then quickly plunged into sadness, crying, "How many people really think about what it's all about? How many people ever, ever think—just once—about love?" As quickly as the grief had overtaken him, laughing returned. He told a joke: "Joe, what do you do when your engine quits at two hundred feet?" He convulsed with laughter at the reply: "You land the sonofabitch!" He sang some more, pondered getting out, and angrily replied to a voice he heard speaking to him: "You voice! Keep quiet up there! Quiet!" He then sighed deeply, felt profoundly bored, and returned to thoughts of the space monkey, commenting with irritation, "I might just as well be Sam, for all I can be or do or think or hear or be or smell or taste!" For ten minutes he convinced himself that his time in the tank was unprofitable because it revealed no information and he felt fine. He questioned again whether he heard noises, then suddenly got out of the tank. During his four and a half hours spent floating in isolation, his longest period of silence was less than six minutes long. In a post-run interview he appeared calm, even happy, but confessed, "I honestly believe, if you put a person in there, just kept him and fed him by vein, he'd just flat out die!"[25]

The mental images the reporter saw and the voices he heard were quite normal. Half of the mentally healthy men and women who slid into the tank reported hallucinations.[26] These hallucinations were not the result of a deranged human being slipping into madness but simply the natural products of a brain deprived of stimulation from the outside world: the sound of human conversation, the feel of cotton socks, the bitter taste of coffee. Shurley put it another way: the wild images that people often envisioned in the tank were a normal part of an abnormal situation.[27] Of course, some people were startled by the images that suddenly appeared in their minds. They vigorously fought them off and tried to resist free association or introspection. Some people reached for the rim of the tank, hoping to actually feel something. Others tried counting to a thousand or forcing their minds to tick off precise seconds, like a clock that they almost willed into reality. Some people resisted the hallucinations so strongly that they turned their anxiety into physical discomfort: wrenched backs, rigid legs, tight shoulders.[28] Subjects who recoiled from disturbing images began to realize that there was something even more frightening than hallucinations. They discovered that what terrified them was not on the outside but inside them. They could not turn off their minds. They could not control the images that raced through their heads. Thinking itself became the most dangerous thing in the world.*

After about two hours floating in the dark, silent tank, subjects would reach another layer of consciousness. "Being alone tended to make a philosopher" out of everyone, Shurley said.[29] Only a handful of college students who volunteered to float in the isolation tank ever reported negative side effects. Most subjects found the floating experience pleasant.† Everyone

---

*It was possible, Shurley, admitted, to "fake to a degree" one's genuine reaction to the isolation tank. Some subjects would never reveal experiences of sudden fright, would not tell the observer of any hallucination they saw. It concerned Shurley that in conducting his experiments, he never could get rid of the universal prejudice in culture that insisted that hallucinations are related to insanity rather than a product of an abnormal situation.
†Shurley's test results showed that 5 percent of college students in the study reported negative side effects (depression, paranoia, delusions), while 60 percent reported that time in the tank was pleasant.

left the experience a changed person. They discovered something in themselves, Dr. Shurley observed, that altered them, changed the way they looked at themselves and the world. Even days or months after the experiment, they found themselves thinking about floating again, trying to remember how it felt and what they saw. Some people even found themselves unusually encouraged or emboldened after the time in the tank. They began to think about taking risks that they had never considered before. Several people who had been afraid of water before their time in the tank afterward learned to swim.[30]

Jerrie Cobb's run in the tank shattered every previous record. Several hundred subjects had already participated in the experiment, and six hours in the water was thought to be the absolute limit of tolerance. Cobb remained in sensory isolation for nine hours and forty minutes, her run finally terminated by Walters. Shurley was astonished by Cobb's ability to withstand solitude. Equally extraordinary was the taped transcript of what Cobb said in the tank. The reporter's monologue during his four-and-a-half-hour run had gone on for pages and pages; Cobb's comments during a period more than twice as long filled only two sheets of paper.[31]

From the moment Cobb stepped into the tank until the time she wrapped a robe around her wet shoulders and walked back into the observation room, she was a model of peace and control. Nearly all of Cobb's comments in the tank were objective observations of external stimuli or, more precisely, the absence of such sensation. Like an astute observer of the natural world, she indicated the position of her feet and the lack of differentiation between her body and the water. Only one time did she mention physical discomfort, when she was forced to move slightly in order to work out a crick in her back. Her only comments of a subjective personal nature were brief and indicated that she felt calm and peaceful. "Just reporting again, everything's fine in here . . . peaceful and quiet and restful," she said. Although Dr. Shurley and Walters had been careful not to give Cobb any indication of how she should respond, Cobb had already decided that success meant remaining as quiet and calm as possible. Writing in her autobiography several years later, Cobb noted that before her run she had heard of other subjects who had reacted to sensory isolation with unexpected, surprising, and often embarrassing responses. She hoped to avoid that. When one brief dream in the tank conjured up images of Cobb's black dachshund, her sister, and a hot summer day, Cobb chose not to report it into the microphone. Seemingly chagrined by the illogical randomness of the dream, Cobb recalled the images as "ridiculous" and later prided herself on not succumbing to any imaginary fantasies.[32]

Cobb's frequent comments on feeling peaceful and relaxed were, of course, consistent with the reality of the tank; for most people it was a calm environment. In Shurley's later analysis, however, he identified a bit of "the lady doth protest too much." That unintentional revelation occurred near the end of Cobb's run when she offered "everything's fine in here . . . and I think I'll get out of the tank unless you want me to stay in longer. . . . I . . . I don't think my feelings are going to change by staying in here any longer so I don't see any need in doing so, but I'm perfectly happy and willing to and I just feel real calm and relaxed." Shurley viewed Cobb's statement as peremptory, suggesting that she feared her feelings soon would indeed change and she wanted to be ahead of the game. Shortly after that comment Cobb reported that she saw a dim light in the tank room, one that looked as if it were "creeping in from under the door all the way across." Shurley believed Cobb started to hallucinate at that precise moment. When the lights finally came up, Cobb confessed that the light she thought she saw had not actually been there. Shurley found Cobb's suggestion that her feelings would not change indicative of a subject who was afraid to abandon herself to a hallucination. In fact, it sounded quite like the reporter who, at the end of his run, stated that he believed his time in the tank was not producing any useful scientific data.[33]

What interested Dr. Shurley even more than her insistence on a peaceful world were Cobb's repeated references to the pleasures of immobility. Three times in her meager monologue she mentioned preferring inertia: "I find the less I move the more I like it"; "But I dislike moving, I'd rather be real still"; "I dislike moving." In Shurley's later analysis, Cobb's spontaneous attention to immobility revealed more than simply preferring to remain stationary. He believed her comments conveyed a deeper desire to oppose movement of any sort, including mental movement. To him, Jerrie Cobb was an enigma. She was a careful pilot whose safety depended on her ability to accurately interpret data around her. While in a cockpit, she displayed a keen awareness of weather, wind, the feel of her plane, and a thousand other external details. On the ground, she was a woman who viewed the world through rose-

colored glasses and exerted considerable mental energy to see human inter-
actions in a positive light. He found her "remarkably unable to allow herself
the freedom of her own mind, the freedom of thinking whatever she wants to
think." The desire to remain fixed also included avoiding forms of alternative
reality that she could not verify or did not desire: hallucinations, mental
images, memories, dreams. She saw what she wanted to see. Several weeks
later Dr. Shurley was startled to discover that Cobb had invited her friend
Jane Rieker from *Life* magazine to stage infrared photographs of Cobb reen-
acting her time in the tank. "Damp Prelude to Space," *Life* entitled its two-
page spread, which was published several months later. In a way that he
hadn't quite expected, Shurley confessed, Cobb wanted publicity for her
feat.[34] Knowing she had surpassed every previous record in the tank, Cobb
hoped the *Life* magazine publicity would demonstrate to the world her mental toughness.

In a medical report to Randy Lovelace, Dr. Shurley summarized his findings on the full range of psychological tests Cobb had taken in Oklahoma
City and offered clinical predictions regarding her potential as an astronaut.
Clearly he was impressed with Cobb's mental abilities and noted in particular
her high level of motivation for becoming an astronaut. In fact, Jerrie Cobb
seemed to sublimate most of her basic drives into her aviation work, he
reported. Flying was everything to her, and he deemed her a "healthy, functioning action-oriented personality." Dr. Shurley also drew attention to
Cobb's laconic communication style, observing that while she was coopera-
tive and goal-directed, she volunteered scant information and shied away
from elaboration. "One is struck," he wrote, "by her economy of both words
and motion." Shurley then summarized the results of her first two days of
psychological tests, which revealed that her intelligence was in the bright-
normal range, she displayed high psychomotor coordination, engaged in
minimal introspection, and was more of a "doer" than a creative thinker.
Cobb appeared a "highly conforming, moral, and conventional person," he
wrote. Cobb's sensory isolation test received extensive comment. She "far
exceeds any other female yet tested" and revealed extraordinarily effective
psychophysiological functioning and adjustment. Shurley noted her peaceful attitude and her lack of perceived stimuli, adding that she was "somewhat constricted" in what she reported. In his conclusion, Shurley stated
that he believed Cobb possessed exceptional if not unique qualities for serv-

Jacqueline Cochran gives instructions to WASPs at Camp Davis, North Carolina, around 1943. *Courtesy The Woman's Collection, Texas Woman's University.*

Dr. W. Randolph Lovelace II collapsed on the ground in Ephrata, Washington, after his famous parachute jump in 1943; at the time, experts believed his descent from eight miles up was the highest ever attempted in the United States and quite possibly the world. It was Dr. Lovelace's first and last parachute jump. *Copyright © The Boeing Company.*

The Lovelace Foundation and Clinic, in Albuquerque, New Mexico, provided secret testing for Project Mercury astronauts and helped select the final seven men who became the first Americans in space. *Lovelace Health Systems.*

Jerrie Cobb consulting with Dr. Lovelace at the Lovelace Foundation, 1960. When Lovelace publicly presented the results of Cobb's successful astronaut tests at a medical symposium in Stockholm, a media frenzy erupted. Within hours, Cobb's photograph was in newspapers across the world as the United States' first "astronette." *Ralph Crane/TimePix.*

Air Force Brigadier General Donald Flickinger hoped to begin a "girl astronaut" program at Wright-Patterson Air Force Base, Dayton, Ohio. When military brass protested, the experiment ended in 1959, and Flickinger handed it to Dr. Lovelace. *National Archives.*

NASA Administrator James Webb swearing in Jacqueline Cochran as a space agency consultant, 1963. *NASA.*

Ruth Nichols, a pioneer in women's aviation, and holder of speed, altitude, and nonstop long-distance flying records. Nichols reported that at age fifty-eight she "sampled" some of the astronaut tests at Wright-Patterson Air Force Base.
Courtesy Nancy H. Greep.

Jerrie Cobb during her 1959 jet flight. Cobb, like other accomplished women pilots, was rarely offered the opportunity to fly military jets. When allowed, Cobb was able to fly only as an "official visitor" and provided she was accompanied by a male pilot. *NASA.*

Twenty-two-year-old Wally Funk at Fort Sill, Oklahoma, in 1961, where she taught military personnel how to fly. Funk read about Jerrie Cobb in *Life* magazine and immediately wrote Dr. Lovelace and volunteered for astronaut testing.   *Courtesy Wally Funk.*

Jean Hixson and students at Crouse Elementary School, Akron, Ohio, 1951. When she broke the sound barrier in 1957 in an Air Force jet, her students were awestruck and she became known as the "supersonic schoolmarm."
Courtesy Pauline Vincent.

Irene Leverton preparing for the International Pendleton Air Races in 1965. Leverton organized her own women's pylon air races and placed first and second two years in a row.   *Courtesy Irene Leverton.*

Gene Nora Stumbough was forced to quit her job teaching flying in order to be ready for additional astronaut testing at the U.S. School of Aviation Medicine in Pensacola. When the Pensacola testing was abruptly canceled, Stumbough was out of work but eventually landed what she called her "dream job," flying as a demonstration pilot for Beech Aircraft in 1964. *Raytheon Aircraft.*

Bernice "B" Steadman won all the major women's air races at least once, including the 1966 Powder Puff Derby. She later joined a long line of extraordinary women pilots, including Amelia Earhart and Jacqueline Cochran, who served as president of the Ninety-Nines, the international organization of women pilots. *Joe Clark/TimePix.*

After the Mercury 13 were denied further astronaut testing, Rhea Hurrle Allison moved to Colorado Springs, where she occasionally towed gliders for the Air Force Academy. Male cadets never knew that the pilot towing their gliders in the early 1960s hoped to be among the first women in space. *Courtesy Rhea Woltman.*

Jerri Sloan greeting her son David after competing in the 1961 Dallas Doll Derby with Martha Ann Reading. Sloan's week of astronaut testing at the Lovelace Foundation was complicated by her pilot husband's jealousy and angry calls to Dr. Lovelace. *Courtesy Jerri Truhill.*

Jan (left) and Marion Dietrich in 1956. Known in racing circles as "the twins in a twin," the Dietrich sisters enjoyed Jacqueline Cochran's preferential treatment during their testing at the Lovelace Foundation. *Courtesy Patricia Daly.*

Air Force Reserves officer Jean Hixson at Wright-Patterson Air Force Base, around 1960. Hixson had exceptional experience as a pilot, including high-altitude flying, but lied about her age when applying for astronaut testing. At thirty-seven, Hixson shaved off two years, unaware that Dr. Lovelace had already decided to test even older women. *Courtesy Pauline Vincent.*

Jerrie Cobb completing an endurance test at the Lovelace Foundation in 1960. A former international ferry pilot and aviation world-record holder, Cobb grew impatient with reporters' questions that focused more on prospective boyfriends and favorite recipes than her determination to succeed as an astronaut. *Ralph Crane/TimePix.*

Jerrie Cobb on tilt-table test at the Lovelace Foundation in 1960. Scientists pronounced Cobb an exceptional human specimen, equal in physical fitness and ability to endure the stresses of space flight to the Project Mercury astronauts. *Ralph Crane/TimePix.*

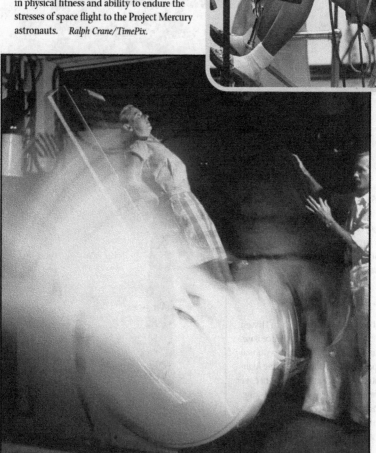

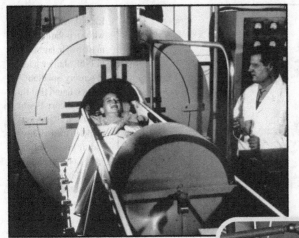

Jerrie Cobb undergoing total body count at Los Alamos Scientific Laboratory, New Mexico, in 1960. Cobb likened the experience to being placed in an iron lung.
Lovelace Health Systems.

Myrtle Cagle pushed to exhaustion on the Lovelace bicycle endurance test in 1961. Cagle contacted Jacqueline Cochran about participating in the program and looked to Cochran, not Cobb, as the leader of the Mercury 13.
Lovelace Health Systems.

Jerrie Cobb performing maneuvers on the MASTIF at Lewis Research Center in Cleveland. The NASA facility opened its doors to Cobb's testing in 1961 and provided her with a dizzying test run on an enormous gyroscope designed to simulate movements in an out-of-control spacecraft. *NASA.*

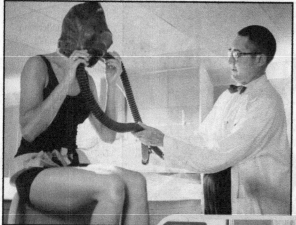

Rhea Hurrle being fitted with an isolation tank mask by Dr. Jay Shurley in 1961. Shurley, a leader in sensory isolation studies, found that three of the Mercury 13 women surpassed all previous marks in their ability to endure intense solitude. Jerrie Cobb, in particular, struck Shurley as "a girl who excels in loneliness." *Carl Iwasaki/TimePix.*

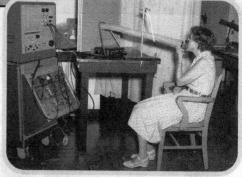

Jean Hixson undergoing pulmonary tests at the Lovelace Foundation in 1961. *Lovelace Health Systems.*

Five days after Sarah Gorelick's farewell party at AT&T, she received a telegram from Lovelace saying the Pensacola tests had been canceled. *Courtesy Sarah Ratley.*

The Mount Holyoke College Class of 1940 protesting NASA's prohibition on women astronauts during an annual alumnae reunion parade in 1965.
Photo by Vincent D'Addario; The Mount Holyoke College Archives and Special Collections.

John Glenn, Annie Glenn, and Vice President Lyndon Johnson in a parade to the United States Capitol following Glenn's orbit of the earth in 1962. Later that year, Glenn's testimony before Congress helped end the Mercury 13's chances of becoming astronauts.　　© *Bettmann/CORBIS.*

Editorial cartoons such as this one from 1962 portrayed Vice President Johnson as hounded by frivolous women who wanted to become astronauts. *The Dallas Morning News.*

Editorial cartoons at the time of Sally Ride's 1983 space flight captured much of the same public sentiment concerning women astronauts expressed more than twenty years earlier.
Reprinted with special permission of King Features Syndicate.

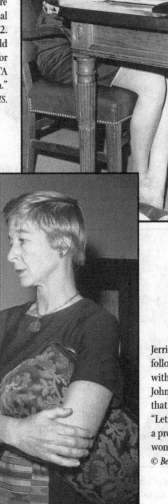

Jerrie Cobb and Jane Hart testifying before the congressional subcommittee in 1962. "Let's face it," Hart told the subcommittee, "for many women the PTA just is not enough."
© *Bettmann/CORBIS*.

Jerrie Cobb and Jane Hart following their meeting with Vice President Lyndon Johnson in 1962, unaware that Johnson had written "Lets Stop This Now!" on a proposal to consider women astronauts.
© *Bettmann/CORBIS*.

Sally Ride repairing the air filtration system aboard the space shuttle *Challenger* in 1983. Ride became the first American woman in space twenty years after the Soviets launched Valentina Tereshkova.   *NASA.*

The first class of women astronauts, 1978: (left to right) Sally Ride, Judith Resnik, Anna Fisher, Kathryn Sullivan, Margaret Seddon; not pictured, Shannon Lucid.   *NASA.*

Wally Funk (upside down) performing weightlessness exercises while participating in a space adventure program in Star City, Russia, in 2000.
Courtesy Wally Funk.

Mercury 13 women attending a 1995 space shuttle launch: (left to right) Gene Nora Stumbough Jessen, Wally Funk, Jerrie Cobb, Jerri Sloan Truhill, Sarah Gorelick Ratley, Myrtle Cagle, Bernice "B" Steadman. *NASA.*

Eileen Collins in "left seat," training in 1998 for a space shuttle flight. Collins was the first American woman to command the shuttle and argues that her accomplishment would not have been possible without the early efforts of the Mercury 13. *NASA.*

ing as an astronaut, citing her ready acceptance of direction and responsibility and noted the "unusually smooth integration" of her body and mind.[35] To Cobb's friend Ivy Coffey, who later interviewed Dr. Shurley for an article on Cobb's run in the sensory isolation tank, the psychiatrist offered his most revealing analysis. Shedding professional medical jargon, Shurley went right to the point. Jerrie Cobb, he said, is a "girl who excels in loneliness."[36]

Cobb's psychiatric testing in Oklahoma City proved so useful in adding to her credentials as an astronaut candidate that she contacted the other Mercury 13 women and offered the phase two experience to them. "Dr. Lovelace and I have discussed the importance of the testing for all women in the program and hope that you will be able to participate," she wrote. Financial assistance to cover the cost of traveling to Oklahoma was not available, however. Cobb did not want to involve Jackie Cochran in the additional testing, but she realized that many of the women would not be able to afford a quick trip to Oklahoma City. She offered to let the women stay at the home she shared with Ivy Coffey. Since Cobb's invitation came just as Randy Lovelace's plans for spaceflight simulation tests were shaping up, most of the women decided to wait until Lovelace's plans were announced before making arrangements to take the isolation tests in Oklahoma City. The wording of Cobb's letter—a letter she addressed to the "F.L.A.T.s" (Fellow Lady Astronaut Trainees)—also implied that taking the isolation tests, while a good idea, was not required.[37]

Only Rhea Hurrle and Wally Funk could go to the Oklahoma City Veterans Hospital right away. Hurrle arrived first and stayed in Coffey and Cobb's guest room. In a playful mood, Cobb bought spaceship-printed bedspreads, wallpapered the ceiling with stars and views of the planets, and added signs saying "Have urge, will orbit" to what she called the "space dormitory." Cobb was impressed when she finally met Rhea Hurrle and found out that Hurrle's employer had loaned her a Piper Comanche to fly from Houston. Cobb thought the gesture spoke well of the young woman's flying ability.[38] Wally Funk, always eager to prove her abilities, arrived on Hurrle's last day of testing. Over a backyard barbecue the three women compared experiences at Lovelace and listened as Cobb described what she knew about the next steps of the program and her hope for the future. They exchanged information they had about the other women who had passed the Lovelace tests. After such a long period of secrecy and isolation, the chance to become acquainted was welcomed. The women were excited and full of energy. Not only did they begin to feel part of a group, they also seized the opportunity to size up the competition. That was certainly the case when Cobb asked the women to join her in physical fitness drills in the backyard. If it was not clear to Hurrle and Funk who was the leader among the women astronaut candidates, they learned quickly when Cobb gave the signal for sit-ups.[39]

The time Rhea Hurrle and Wally Funk spent in the tank and their responses to sensory isolation surpassed Cobb's extraordinary run. Hurrle spent ten hours in the tank before her run was stopped by the testing observer. She exhibited minimal movement, spoke little, offered only objective remarks, minimized physical discomfort, and revealed no perceptional distortions or hallucinations. No manifest anxiety, spatial disorientation, or difficulties adjusting to simulated weightlessness were apparent. In other tests she displayed what Shurley called strong motivation for spaceflight and high tolerance for stress. Wally Funk also excelled in the tank. Demonstrating the sheer persistence she had shown on the bicycle test at Lovelace, Funk remained in the tank until observers asked her to come out. In the report to Dr. Lovelace, Shurley wrote that Funk "gave no evidence of any approach to limits to her tolerance. She impressed instead that she could and would have continued indefinitely in the situation in absence of 'outside' intervention." Funk's total time in sensory isolation was ten hours and thirty minutes. During that time she uttered not a single word.[40]

Expecting the other Mercury 13 women to arrive at his laboratory over the summer, Dr. Shurley drew up a standard research budget, which he was required to submit to the Veterans Administration for approval. The budget outlined the expenses involved in testing each woman. Almost all the expense were personnel hours: time spent doing mental status examinations and write-ups, conducting psychological testing and physical exams, a research assistant's time spent maintaining the equipment in the sensory isolation lab, making and transcribing audiotapes, doing routine lab work for blood and urinalyses. Shurley estimated the testing would cost the VA $556.39 per woman candidate and could run as high as nearly $7,000 for the entire group. Shurley expected to see at least eight of the women over the summer of 1961, although he hoped eventually to test all the candidates. Conversations among Cobb, Shurley, and Dr. Lovelace in Albuquerque also suggested the possibility of involving the Project Mercury men in tank runs in order to calculate the differences in response between men and women. As he prepared for additional testing, Shurley pulled out a nondescript office folder into which he started filing reports, notes, and observations for the women-in-space program. Realizing that Randy Lovelace had not supplied him with a name for the program, Shurley thought for a moment and then penned his own. "Project Venus," he wrote, and then tucked the folder into a filing cabinet.[41]

CHAPTER EIGHT

Waiting for Pensacola

RANDY LOVELACE FOUND A SITE FOR "PHASE THREE" SPACEFLIGHT simulation tests. The U.S. Naval School of Aviation Medicine consented to Jerrie Cobb's undergoing ten days of testing at its facilities in Pensacola, Florida. If Cobb's tests went well, the rest of the Mercury 13 would follow a few weeks later. Dr. Lovelace wrote each of the women and asked them to make arrangements for time off in June. Jerri Sloan asked her mother to come to Dallas and help her with child care. Sarah Gorelick worried about how she could ask for additional days off from her engineering job at AT&T. Janey Hart was thankful the Senate was in recess and her husband could help more with their children. As the women cleared their schedules, Lovelace wrote again and said the date had been moved to sometime in July. Like the tests at Albuquerque, the Pensacola exams were squeezed in around previously scheduled work. While Randy Lovelace was grateful that the Navy would support the women's astronaut testing, the on-again, off-again scheduling made finding baby-sitters and talking to unsympathetic bosses even more difficult for the women.

Jerrie Cobb was lucky to have a fixed date for phase three testing, and she was equally fortunate that her boss at Aero Design was always willing to give her time off for astronaut business. In fact, Tom Harris, who had introduced Cobb to Don Flickinger and Randy Lovelace on the beach in Miami two years earlier, had extended himself even further to help Cobb's cause. On several occasions, when aviation business took him to Washington, he tried unsuccessfully to talk with NASA's chief, James Webb, a friend since Webb's Oklahoma days. He also lobbied Oklahoma senators. Tom Harris wanted to see Cobb given a chance to become an astronaut and he was concerned about the country beating the Russians in at least one space race. As a businessman with an international perspective and a keen interest in politics, he also wanted to play a part in national decision making. He was already talking with state Republican Party bosses and making known his interest in running for the U.S. Senate. With Harris's political contacts, Janey Hart's influence through her husband, Senator Philip Hart, and Jerri Sloan's mother, who knew Lyndon Johnson from working on his Texas campaigns, Cobb had persuasive allies for advancing the cause of women astronauts.*

Before leaving for Pensacola, Cobb spent several days training for the tests at the home of *Life* magazine writer Jane Rieker in Delray Beach. Not knowing exactly what physical endurance challenges she would face, Cobb trained by running, swimming, and becoming acquainted with the effects of simulated weightlessness. A neighbor of Rieker had a swimming pool he was willing to share and Cobb spent hours below the surface in scuba gear. Every day, oxygen tanks were delivered for her scuba diving, and the metal containers sat lined up along the pool like military sentries. Often Cobb would remain so motionless at the bottom of the pool that the only evidence of her presence was small bubbles gurgling to the top of the water. On several occasions, Rieker's neighbor became anxious about Cobb's immobile marathons and darted out of the house to peer into the water to make sure she was still breathing.[1]

Cobb reported to the Pensacola base on a hot, muggy May evening and stowed her gear in the officers' quarters. After uneasy sleep in the noisy barracks, she reported at 8 A.M. for the first of her tests. Unlike the previous exams, which had evaluated her physical and mental condition, the next ten days hurled her into the realm of the hypothetical with tests that would determine if she could handle the stresses, confusion, and physical assault of outer space. These exams would judge her reaction to space-high altitudes, high G loads, motion sickness, violent seat ejection, and even the turbulent

---

*Thomas Harris ran in the 1963 Oklahoma special election primary to fill the term of Senator Robert S. Kerr, who died in office. Thomas Harris lost the Republican primary to University of Oklahoma football coach Bud Wilkinson. Democrat Fred Harris later defeated Wilkinson in the Oklahoma Senate race.

chaos of trying to escape from a submerged cockpit—an underwater test that gave Cobb a frightening feel for what it would be like to be trapped inside a Mercury capsule swirling to the bottom of the Atlantic. If she passed the tests, she would prove once more that she had the same physical ability and mental determination that the Project Mercury astronauts exhibited. Combined with her successful completion of the physical exams at the Lovelace Foundation and psychological and psychiatric evaluations in Oklahoma City, the Pensacola tests would allow her to assert that she had passed all the tests the Project Mercury astronauts had taken. If some of the tests she took were not exactly the same as those the men took at the Wright Aeromedical Laboratory, they were certainly comparable and, in the case of the sensory isolation test, more rigorous.*

After preliminary tests to bring her medical profile up to date, Cobb launched into a day of physical fitness drills that were used to evaluate the strength and ability of healthy Navy aviators. Scaling a tall concrete wall became a none-too-subtle metaphor for the rest of the week. The wall was six feet, six inches high and built to pose a challenge to men who were asked to leap, grab hold of the top, and throw themselves over to the other side. On her first attempt, Cobb jumped, failed to get a grip, and fell to the ground. On her second try, she ran harder, leaped higher, and clawed her way to the top.[2] At five feet, seven inches in height, Cobb surmounted a wall that was higher for her than it was for men who were, on average, several inches taller. It did not, however, seem to matter. Sit-ups, pull-ups, marathon runs in the Florida heat—Cobb tackled every test and succeeded each time.

For the rest of the week, Cobb took part in spaceflight simulation tests. In the high-altitude chamber, she had her first opportunity to wear a full pressure suit. Just as Betty Skelton had discovered a year earlier when she

*After Project Mercury astronauts' military and aviation records were reviewed, they completed three phases of testing. The first was medical testing at the Lovelace Clinic. The second and third were completed at Wright-Patterson Air Force Base in the Aeromedical Laboratory and included psychological and psychiatric testing described on pages 104–106. The Wright-Patterson tests also included stress tests that involved pressure suit tests, treadmill tests, the Harvard Step Test (subjects step up onto a platform twenty inches high every two seconds for five minutes to demonstrate their physical fitness), acceleration tests, vibration, heat, noise equilibrium, and tilt-table tests, cold-pressor tests, and a complex behavior simulator in which candidates demonstrated their ability to react in confusing situations by facing a panel of twelve signals, each requiring a different response.

demonstrated astronaut tests for the *Look* feature, the military did not design clothing with a woman's body in mind. When Cobb tried on the smallest pressure suit the Navy could locate, it was still too large. Navy personnel spent an hour and a half sealing and strapping her into the bulky encumbrance and then escorted her into the altitude chamber, where technicians ran her up to 60,000 feet and watched closely to see if she could retain her mental acuity and move her legs and arms against the heavy pressure. With her hands encased in ill-fitting gloves that swelled into balloons, Cobb struggled to make a fist and touch her thumb to each finger, demonstrating that she could manipulate sticks, knobs, dials, and latches. To measure Cobb's ability to withstand a rapid high-altitude descent, technicians altered the pressure of the room and brought her down to sea level in a free fall.[3]

Another test required that Cobb sit in the copilot's seat of a Douglas Skyraider as the pilot took the plane up and swung through a series of stomach-turning aerobatics. Eighteen needles were wired to Cobb's scalp in order to record an airborne electroencephalogram of brain activity—an experience she had never encountered before. A camera positioned directly in front of her face caught her barely detectable flinches as the plane dove, looped, and sliced through the sky. With each sudden move of the plane, the force slammed Cobb against the seat, whipped her to the side, and flung her forward. Staring straight ahead, Cobb blinked twice. Her eyes pushed forward against their sockets and then sank back. "Eyeballs out, eyeballs in," pilots called it.[4]

The Multi-Place Ditching Trainer, or what Navy airmen termed the "Dilbert Dunker," tested Cobb's ability to withstand the disorientation of a landing on water. For pilots, an ocean ditching was an unusual event, but one for which they needed experience. Practice on the "Dunker"—a harrowing, chaotic experience—might save the life of a military pilot shot down over water. For U.S. astronauts, however, a splashdown was the only method of landing a space capsule. To the unfamiliar eye, the "Dilbert Dunker" looked like a backyard contraption rigged up by teenagers with creative minds and too much free time. What appeared to be a big round oil drum sat on top of a steep track that ran into a swimming pool sixteen feet deep. The odd-looking craft took on a more serious form once it was propelled into motion. First Cobb squeezed in—it was a tight fit, considering she was wearing a Mae West life preserver and a parachute pack, not standard astronaut wear but bulky enough to feel like a space suit. Then she buckled into her seat harness, set her jaw, and waited. With a jolt, the drum shot forward, hurtling down the track until it crashed into the pool. Instinctively, Cobb held her breath as the craft turned upside down and water flooded in. She tried to remember what she had been told: don't panic, unbuckle your harness, avoid getting your gear looped around hooks, search for a reference object—a latch, an armrest, the bottom of a seat. Make your way to the hatch, slide out, and bob to the top of the water. Cobb made it out of the contraption without the help of rescue divers and in the requisite amount of time. Adrenaline pumping, she heaved herself out of the pool and tried to catch her breath.[5]

One afternoon right after lunch, Cobb faced another aerodynamic carnival ride. It was a slow-rotation room, a windowless enclosure constructed to resemble an apartment with furniture, a bedroom area, a washstand, even a hot plate. But unlike a standard room, the structure was perched atop a forty-two-ton steel gyroscope that spun around at ten revolutions per minute. Because the room had no windows, a test subject experienced no sense of speed, only turbulent disorientation. "Just follow instructions," a voice said over the intercom as Cobb sat in front of a maze of dials, knobs, and switches. As the room spun, the voice asked Cobb to flip switches and set dials on command in order to evaluate how accurately she could perform quick maneuvers while in a confusing, whirling environment. Cobb's head spun as she reached up, down, back, forward, left, right with every command. For a moment, she felt her gorge rise, so she tried to narrow her field of vision and concentrate on each single task. Gradually the sickening swirls in her head eased and the queasiness subsided. Just as she was getting used to the motion of setting dials, the voice asked her to stand up and walk across the room to a dart board. With the room still slowly rotating, Cobb's senses were so upended that she was unaware that the wall had tilted to a wild angle. When the voice asked Cobb to toss a dart at the board, she completely missed the target. Then she realized she had to compensate for the uneven floor by aiming forty-five degrees to the left. When she was asked to pitch tennis balls into a wastebasket, Cobb used the same technique and threw far to the left, hitting the target but operating against what her eyes told her to do.[6]

By the end of her ten days at the School of Aviation Medicine, Cobb was delighted to hear confirmed what she already sensed was true. She had passed phase three space simulation tests, scoring as well as experienced Navy pilots. Pensacola military personnel threw her a party on the base to announce the news.[7] Randy Lovelace sent the rest of the Mercury 13 a letter informing them that their testing date had now been set for July 18.[8] Again, the women scrambled to make arrangements for ten days in Pensacola. Buoyed by her performance, Cobb took an uncharacteristic step and wrote NASA Administrator James Webb directly, telling him about her Pensacola experience. Included in her letter was a newspaper article she called disturbing: a report detailing Russia's interest in sending a woman into space. If an American woman were allowed to launch, the feat "would have tremendous prestige value and also prove our systems as reliable and safe," she argued. She closed her letter by impressing upon Webb her willingness to take whatever time it took to prepare for a spaceflight. After spending one and a half years being tested, proving herself again and again, she told Webb that she would gladly wait ten more years if it meant she could eventually go into space.[9] Cobb hoped that she would have an opportunity soon to speak personally with Webb as both of them were headed to Tulsa, Oklahoma, for a meeting on peaceful uses of space.

The Tulsa conference was an impressive gathering, filled with NASA officials, aerospace contractors, politicians, and members of the media. Randy Lovelace, in one of the main addresses before the group, discussed challenges facing the space program and spoke publicly for the first time about the group of twelve women pilots who had completed phase one astronaut testing. Just as the first announcement of Cobb's testing had created a swell of excitement, Lovelace's revelation energized the crowd. Once more, attention focused on Jerrie Cobb as reporters fired off questions about the possibility of an American woman in space. Cobb, always the proper professional woman in white gloves, obliged photographers by standing next to a mock-up of the Mercury capsule. Her smile was strained as she posed looking more like an attractive model for the latest Chevrolet than a prospective astronaut. That evening at the conference banquet, Cobb found herself seated next to James Webb on the dais overlooking two thousand people.[10] They barely had time to talk, much less discuss Cobb's recent letter, since Webb had to eat his meal quickly before delivering the evening's keynote remarks. Toward the end of his speech, which focused on the triumph of Alan Shepard's launch and the challenges ahead, Webb introduced Cobb to the audience and praised her accomplishments in passing three phases of testing and her dedication to the space program. Webb knew there was national interest in sending a woman into space. So many women had written to NASA asking how they might become astronauts that the space agency had developed a form-letter response. The letter was a polite, inflexible rebuff. The greatest source of astronaut manpower lies with test pilots, the rejection read. When we expand opportunities, "perhaps some inspired offers such as yours can be accepted."[11] As Webb came to the conclusion of his Tulsa speech, he seemed to move into extemporaneous remarks. Without any warning to Cobb or Lovelace, the NASA administrator made a startling announcement. At that moment, Webb said, he was proud to appoint Jerrie Cobb a special consultant to the National Aeronautics and Space Administration.

Cobb was ecstatic, even dumbfounded, but did not have time to talk any further with Webb about what he had in mind concerning her appointment. She was due in Paris for the air show and annual meeting of the Fédération Aéronautique Internationale and left for Europe immediately. Although Cobb was not certain what Webb wanted her to do, she assumed her work would focus on women in space. The vagueness of her charge did not matter to Cobb, nor did she wonder if the appointment was a manipulative way for NASA to co-opt her, bringing her into the fold so they could control her comments on women astronauts.[12] Cobb saw what she wanted to see and later wrote that things had never looked rosier. In a letter from Paris, Cobb told Webb, "I am truly humbled by the confidence you have placed in me and will do my utmost to accomplish any assignments you may give me."[13]

Jackie Cochran was not so pleased. As president of the FAI, Cochran was directing the Paris aviation meeting and had heard about Cobb's exceptional tests scores in Pensacola. Learning that Cobb was now working for NASA made Cochran angry. She resented Cobb's good fortune and access to power and thought her appointment was ill-conceived. Cochran was incensed as well because she had learned about the Pensacola testing from one of the other women astronaut candidates, not from Randy Lovelace. She was beginning to wonder if Lovelace was purposely keeping her uninformed. Cochran had always prided herself on being among the elite circles of aviation. She knew that reporters looked to her for an insider's news and privileged information. Now when journalists called and asked her about the women's astronaut program, Cochran found herself stumbling over answers, unable to offer much more than vague generalities. She did not know which irritated her more: hearing secondhand news about the women-in-space program or appearing uninformed before the press.

Cochran decided she would speak directly to Jerrie Cobb and asked her secretary to contact Cobb, who, they discovered, was staying in their same Place de la Concorde hotel. Their conversation never happened, following the pattern the women had established that spring. Once more, Cochran proposed getting together and, once more, Cobb politely declined. Cobb said she received Cochran's invitation too late to accept. She had already gone to dinner and returned to the hotel far too late to call. Cobb left a note for Cochran at the hotel reception desk, a convenient way to avoid actually speaking with her.[14] Cochran called her husband, Floyd Odlum, back in California and boiled over.

That afternoon Odlum wrote Lovelace a long letter which he marked "personal and confidential." "Jackie does not want to be around if you don't want her," Odlum wrote, and spelled out the reasons Cochran believed Lovelace was intentionally limiting her involvement. The list of long-simmering accusations spewed forth: Why did you start the women-in-space program without consulting with Jackie in the first place? Why did you keep Jackie in a "detached" position when she came to Albuquerque for the Dietrichs' testing? Why was the original list of women candidates prepared without Jackie's involvement? Why have you shown a lack of initiative in getting Jackie lined up with NASA? Why have you become so involved with *Life* magazine in publicizing the program? Why is "a certain Air Force General who is prominent in air and space medicine" trying to push Jackie out of the program? Odlum's final swipe tore into Lovelace's staff at the foundation and what he called their "self-contained and self-sufficient attitude." In concluding, he declared, "If you have personal or other problems dealing with the point of this personal letter you should lay them on the table because, as I said above, Jackie is rather unhappy."[15]

Lovelace wasted no time in responding, and even though he understood that Floyd Odlum's influence and financial assistance were invaluable to the Lovelace Foundation, he did not cower. Nor did he respond to each of Odlum's accusations, focusing instead on the current state of the program without addressing what led up to the Lovelace testing or what might occur in the future. When the women-in-space program was started, he explained, Cochran had been in Europe and unavailable for assistance. There was no special arrangement for publicity with *Life* magazine, although the Navy would prefer to deal with *Life* should additional articles appear because the magazine had a large circulation. No Air Force general had played a part in the current testing, Lovelace declared, avoiding any reference to the nucleus of the women-in-space program that Lovelace had inherited from General Flickinger two years earlier. Suggesting he needed to act quickly in the face of constantly changing circumstances, Lovelace pointed out that the phase three Pensacola tests had only recently been approved by the Navy. He offered to postpone the testing one more time if Cochran wanted to be there in person to become acquainted with the women and observe their testing. Lovelace reserved his most direct statements for describing the program's affiliation with NASA and what he knew to be the space agency's official view of testing women. "I had absolutely nothing to do with Miss Cobb's contacts with them [NASA]. Mr. James Webb, the administrator, is from Oklahoma City and had met her though Senator Kerr. I was as much surprised as anyone when Mr. Webb stated that Miss Cobb was to be a consultant to him. As you know, the previous position of the NASA had been 100% against any examination procedures for girls."[16] Lovelace noted that he had tried to reach Cochran in Paris so that they could speak directly but he was unable to make contact with her. But as much as he wanted to keep the women-in-space program moving without delay and with his own hand directing its course, he did not want to compromise his personal relationship with Floyd Odlum and Jackie Cochran. He deeply appreciated their friendship and wanted their relationship to continue. Almost as a second thought, Lovelace wrote an awkwardly worded sentence at the bottom of the final page that conveyed his genuine emotion. "I consider Jackie and you the couple I am closest too [sic] of all the couples I know."[17]

A week later, Cochran followed up with a letter of her own to Lovelace. Her tone indicated that the crisis between them had passed. Rather than wanting Lovelace to account for what she perceived as his slights, Cochran wanted to be more involved with the women-in-space program. As someone who was used to being at the center of important talk with powerful people, it tortured her to be—as Odlum put it—"so much on the edges of it."[18] In her letter she focused on the upcoming phase three testing and informed

Lovelace that she would not return from Europe until the middle of September. She hoped that the Pensacola tests could be moved to the fall and urged Lovelace to give the women plenty of notice so that arrangements could be made with their employers. She reminded Lovelace that she would pay for travel and other expenses while the women were at the Navy base and asked how their status and pay would be addressed in subsequent stages of the program. As usual, she closed with references to Jerrie Cobb. "It is apparent that one of the girls has an 'in' and expects to lead the pack. She has stated as much to others who have reported the conversation to me. Favoritism would make the project smell to high heaven," she wrote. Moreover, Cobb's new affiliation with NASA was a mistake. If anyone should coordinate the group of women, she declared, it should be a noncompetitor. Lovelace did not have to ask who Cochran would propose as a coordinator.

Lovelace moved the testing date, writing to each of the Mercury 13 once again and informing them that the date for Pensacola had been changed to September 18. Within weeks, each of the women received overlapping letters from Randy Lovelace, Jerrie Cobb, and Jackie Cochran about Pensacola. As harried as they were scheduling and then rescheduling the trip, none of the women stopped to wonder if the multiple letters indicated a struggle for leadership with the women-in-space program. Cobb wrote first, assuming an authoritative stance, and extended her congratulations for passing phase one astronaut examinations in Albuquerque. She informed the women that she had already taken the Pensacola tests and suggested they prepare for a physical challenge. Offering information on where they would stay, how they should dress, and where they should take their meals, Cobb also enclosed a release form discharging the United States of any legal claim that might ensue from the testing and asked that the forms be returned to her home.*[19] Lovelace wrote next announcing that Jackie Cochran would be in Pensacola for the testing and that she had offered to cover all the candidates' travel expenses. "You can thank her in person," he suggested. He also indicated that he expected a rush of publicity to follow the phrase three tests and advised the women to hold a group meeting to discuss how they wished to handle it. Lovelace added that the male astronauts had acted as a group for

*The form Cobb sent had no indication that it had been generated by the U.S. Naval School of Aviation Medicine. In fact, the school is inaccurately referred to as the U.S. Naval Aviation Medical Center. Most likely, Cobb herself crafted the form.

publicity purposes, and, inferring that there would soon be a similar media onslaught, he recommended that the women follow suit.[20] Cochran followed with a long letter that served as her formal personal introduction to the women. "As you probably know I am not a participant in these medical checks and tests. . . . Some of you may therefore wonder why my great interest and my assistance." She offered a lengthy response reminding them of her role as a special consultant to Dr. Lovelace and her years leading the WASPs. In looking back on World War II, Cochran wrote, she saw a significant distinction between women taking the lead then and now. "There was a manpower shortage to be considered in those war days," she wrote, "and it can hardly be said that there is any such shortage with respect to astronauts." Her view of the program may have caused some confusion in the women's minds as she seesawed between cautioning them that the tests were "purely experimental and in the nature of research" and calling them "candidates" as Randy Lovelace always did. A "properly organized astronaut program for women would be a fine thing," she said. "I would like to help see it come about." Just as Cobb had done, Cochran asked the women to write her directly if they had any questions.[21]

Sarah Gorelick knew she would have trouble taking time off from her engineering job at AT&T to spend ten days in Pensacola. With her vacation time already spent, her entire department working overtime, and a phase three testing date that had bounced around three times, Gorelick did not see how she could ask her boss for another week away. She had not been able to get away for July sensory isolation testing in Oklahoma City and feared she might miss Pensacola altogether. She wrote to Jerrie Cobb and conveyed her concern. Cobb responded quickly and offered Gorelick a letter she hoped might convince her AT&T boss. The letter was businesslike but inflated as Cobb used every angle she considered persuasive. "This serious program is being conducted on a highly scientific level and is of utmost importance. The Pentagon has recently approved the further testing of women pilots who successfully passed the Albuquerque tests, at a military aerospace medical facility for the last two weeks in September. Although it has been necessary to keep this program 'under wraps' as much as possible after the September tests, the results and names and details will be released. *Life* magazine among other news media will carry the stories."[22]

In addition to corresponding with the Mercury 13 about Pensacola,

Jerrie Cobb also was working on an introductory report for James Webb and NASA on women in space. While in Paris, Cobb had been approached by two Russian scientists asking about U.S. plans to put women into orbit. She also knew that a German observatory had recently picked up the sound of a female voice on an audio frequency used for training Russian cosmonauts. The Russians might already be ahead of the Americans. The time for the United States to commit itself to launching the first woman in space had arrived, she wrote. Cobb outlined three objectives for NASA: to make the first woman in space an American, to gather baseline scientific data on what women could contribute in space, and to educate women and young people about the necessity of scientific education and space exploration. In her arguments, Cobb was careful to avoid suggesting that women would make better astronauts than men. Her task was not to "detract from the true pioneering men of space," she wrote, but she believed strongly that one exception needed to be made to the current schedule for U.S. spaceflight: America needed to win the race of launching the first woman into space. "If a woman can do it," Cobb wrote, using a bit of sexist humor, "it *must* be safe and simple!" She vowed to search the country for one woman who met NASA's astronaut requirements, although her biography and credentials were the only ones submitted with the report.[23]

In early August, six weeks before the Mercury 13 began arriving at Pensacola, Jackie Cochran spoke directly to Robert Pirie, the Navy's vice admiral in charge of air operations, about her concerns. Cochran and the admiral had initially talked about the testing when they had shared a car ride together a few days before, but in weaving through heavy traffic, she was worried that her most important point had not been clear. Wanting to make sure Pirie understood how she viewed a women's astronaut program in relation to Project Mercury, Cochran reiterated her ideas and mailed the letter directly to his home. Men came first, Cochran wrote. She was in favor of a space program for women only if it did not impede or interfere with the progress of the men's program. A women's program should only be instituted "at the proper time and in the proper way" she believed, not elaborating on when the time would be right.[24] Before Cochran's discussion with Pirie, the Navy brass in Washington had been briefed on Jerrie Cobb's space simulation testing. They took the Pensacola exams lightly at the time. According to Cobb, when Pensacola officers wired Washington for permission to administer tests to determine the difference between men and women astronauts, the Pentagon flippantly responded, "If you don't know the difference already, we refuse to put money into the project."[25] In September, however, with more women being tested and with Cochran's concern about a proper time and a proper way, Pirie took the matter seriously and wrote James Webb at NASA to make sure the space agency was interested in women astronauts.[26]

By the beginning of the fall, Jackie Cochran's interest in the Pensacola testing made a surprising shift. She would not be coming to observe the women's testing, Cochran informed Lovelace. From the tone of her letter, Cochran seemed to say that she was involved in more important activities and that phase three testing of the Mercury 13 was inconsequential. Between the end of August and the beginning of October, Cochran was at Edwards Air Force Base in California, piloting a Northrop T-38 Talon to nine new women's aviation records. "Miss Cochran has been flying a company-operated T-38," the Northrop press release offered by way of clarification, noting that T-38s were being flown by the Air Force.[27] With Air Force permission to use the facilities at Edwards and a jet from a top aviation corporation, the fifty-five-year-old Cochran was exactly where she wanted to be: in the cockpit breaking records. Writing to Lovelace, Cochran expressed irritation at the idea of traveling to Florida to stand by and watch other women take tests. "I cannot ask the company to have the ground organization stand by and the plane remain idle while I make trips of no particular interest to Northrop," she wrote Lovelace. "Even if I were there," she added, "I don't think there would be much I could do to be helpful, as a spectator."[28]

Cochran's letter voiced new concerns about the program just as it also revealed how little she actually knew about the women who would undergo the tests. The selection process had not been refined, she argued. The women did not have scientific backgrounds or military experience. Cochran did not seem to realize that Flickinger, Lovelace, and Cobb had reviewed the aviation records of more than 782 women pilots in narrowing the field to the 25 women who had been invited to Albuquerque. Nor did she know, for example, that Sarah Gorelick had degrees in mathematics, physics, and chemistry, or that Jean Hixson had already been trained in explosive decompression at Wright-Patterson Air Force Base. She had even forgotten that Hixson had served under her in the WASPs and worked as an engineering test pilot flying

B-25s. As the September 18 date for Pensacola neared, Cochran threw herself into jet training with the full weight of Edwards, the Air Force, Northrop Aviation, and her friend Chuck Yeager behind her. At the end of one run, Yeager was struck by the close relationship Cochran enjoyed with men in military and corporate circles. Writing in his diary, Yeager observed, "At 12,000 she removed the face piece from her pressure suit and made a perfect landing on the lake bed. Norair presented Miss Cochran with one dozen yellow roses, a very tender ending to a wonderful program."[29]

As the Pensacola testing approached, Sarah Gorelick resigned her position in AT&T's engineering department, not wanting to test her boss's limits or ask her already overworked colleagues to step in for her again. Gorelick realized that she might not be able to easily find another job in the technical field, but she also knew that she had to go as far as she could with the women-in-space program. On September 7, Gorelick's coworkers gathered for a farewell party in the drab Kansas City offices of AT&T. She had not been able to keep her astronaut testing a secret after her repeated requests for time off. Her coworkers had been delighted with the news, and the national office of AT&T had even written a brief story in one of the company publications applauding its employee. At the party, Gorelick was presented with a toy rocket and space helmet. Her friends meant the toys as a lighthearted gift, but they were also genuinely proud of Gorelick and her accomplishments. One of her friends had written "S. Gorelick" across the front of the helmet.[30]

Other women were rushing to make last-minute arrangements as well. Jerri Sloan, in the midst of divorce proceedings, had additional family problems. Her son came down with pneumonia and her father suffered a heart attack. She had no time to train for the Pensacola tests. Lifting children, carrying groceries, arguing with an angry spouse, and driving across the state to care for an ailing parent gave her adequate exposure to stress, she figured. In Georgia, Myrtle Cagle planned to enroll at Mercer University, but she was supposed to be in Pensacola on the exact date the Mercer dean demanded she be present for registration in person. After showing him letters from Jackie Cochran and trying to convince him that she really was being asked to take astronaut tests at a U.S. Navy air base, the dean finally consented and allowed Cagle's husband to register for her. Gene Nora Stumbough also was having some trouble taking time off from her flight instructor job at the University of Oklahoma. It had been fine with her boss to travel to

Albuquerque in the summer. College students did not take flying lessons when school was not in session. But September 18 came right at the beginning of the term and he could not afford to be without an instructor. Stumbough, confident that she would find some way to earn a living, told her boss that she was going to Pensacola, and quit.

For Wally Funk, Rhea Hurrle, Janey Hart, Jean Hixson, B Steadman, Irene Leverton, and Marion and Jan Dietrich, getting ready for September 18 meant riding their bicycles, doing push-ups, and studying aviation manuals and books on meteorology. They reread letters from Jerrie Cobb in which she instructed them that meals in the officers' mess would cost between fifty and seventy-five cents; housing in the bachelor officers' quarters would run two dollars a night. They packed cool sleeveless dresses, shorts, swimsuits, and tennis shoes and placed airline tickets to Pensacola on their bedroom dressers.

On September 12, telegrams arrived at their houses. They expected yet another set of instructions or perhaps another irritating postponement. The news from Randy Lovelace was much worse, and the abrupt Western Union prose made the shock even more jarring. "Regret to advise arrangements at Pensacola cancelled. Probably will not be possible to carry out this part of the program."[31] The reports that trickled in from phone calls, letters, and telegrams over the next few days were even more discouraging. Randy Lovelace had received word from the Navy that NASA had no interest in the tests going forward. After Cochran's conversation with Pirie and his letter to James Webb, NASA determined that sending an American woman into space was not a priority. In the bureaucratic language of the military and the government, Webb's deputy wrote back that "NASA does not at this time have a requirement for such a program."[32] Without a requirement, without the federal government's official approval for the use of military time and equipment, the women-in-space program could not move to the next level of testing. Everything was off. On September 18, the day the women were to begin spaceflight simulation exams in Pensacola, Jackie Cochran stepped into a jet at Edwards and set a new record for straight-line distance. No woman had ever gone farther.[33]

CHAPTER NINE

Changing Course

OVERNIGHT THE QUESTION CHANGED. AFTER THE NAVY CANCELED phase three tests at Pensacola, Jerrie Cobb and the Mercury 13 no longer were fighting to prove females capable of spaceflight. They now had to convince NASA that women had a right to be astronauts. A medical and scientific question suddenly became a political one. With this change in question, a change in strategy and battleground developed. Cobb and Randy Lovelace, who had worked so closely together in screening the women candidates and overseeing their Albuquerque testing, began to operate more and more independently of each other. Lovelace opted for a more diplomatic tone and measured pace in trying to change NASA's mind. Cobb, however, was fueled with indignation and disappointment and all but charged up the steps of Capitol Hill. What Jackie Cochran would do was anyone's guess.

Cobb moved quickly upon hearing the word from Pensacola. She flew to Washington, found a cheap hotel, and started meeting with anyone who would see her. Cobb talked to the military and to politicians on Capitol Hill, and had a conversation with NASA's James Webb, whom she described as sympathetic to the women's cause but unwilling to commit himself.[1] No one had a satisfactory answer. Cobb failed to discover why the Navy had approved her testing but balked when twelve more women wanted to go through the same exams. Nor could she ascertain why Washington was so unenthusiastic about sending an American woman into space when Russia was already bragging about a future female cosmonaut. All Cobb could determine was that NASA did not have a "requirement" for women in space. Even the word "requirement" seemed to imply that unless "required," women were dismissed.

Randy Lovelace realized that it had been a mistake to keep James Webb and NASA uninformed about his women-in-space project. He wrote Webb a detailed letter outlining the project over the past year and politely recommending that the remaining Pensacola tests be allowed to go forward. The letter was a skillful dance and revealed Lovelace's considerable talent for persuasion. He never questioned *if* women should be astronauts, he only asked *when* it would happen. While he hoped that his program would go forward, Lovelace was keenly aware that his foundation depended on NASA contracts, and he certainly wanted to keep federal space projects coming to Albuquerque. The letter tried to ameliorate any irritation NASA felt for his stirring up curiosity in women astronauts while also attempting to convert Webb's annoyance into interest. Nearly everyone who ever worked with Lovelace knew he possessed a gift for generating excitement in others. He had been a powerful catalyst for innovative ideas in the past. He hoped he could spark James Webb's imagination as well.

In defining the purpose of his experiment, Lovelace noted that it was not the project's intent to put a woman into space "at an early date" but rather to amass scientific data on women in order to determine how they might be best used in the future. "All these candidates were informed before coming here that there was no astronaut program for women at the present time and that maybe there would not be one for several years," he wrote to Webb. Lovelace went on to emphasize the seriousness and thoroughness of his project, and commended the thirteen women pilots, calling them knowledgeable and highly experienced. In what sounded like a preemptive move, Lovelace stated that he did not want any press about the project, as he believed publicity could be damaging. Distancing himself from the publicity Cobb had previously generated, Lovelace informed Webb that he had asked all the women candidates to avoid the media. He failed to mention that he had participated in both the *Life* and *Parade* magazine articles that Cobb and Cochran had initiated earlier. On one final matter, Lovelace echoed the attitude of Jackie Cochran. He believed that a women-in-space program should not interfere or impede the men's progress.[2] It was an odd statement coming from a man who was known for being in such a rush—with scientific projects, with advancing medicine, with challenging old-school attitudes about what women could do. As impatient as he was, Lovelace was willing to let the men move forward while the women were held back. Lovelace was shrewd enough to realize that restraint could be strategic, and, of course, he assumed that he would have plenty of time.

Randy Lovelace's letter was not the only one on James Webb's desk urging him to consider someone besides white men as astronauts. Respected broadcaster Edward R. Murrow, then director of the U.S. Information Agency, contacted Webb shortly after the women's Pensacola testing was canceled and asked why NASA did not train qualified black men as astronauts. "Why don't we put the first non-white man in space?" Murrow asked. "If your boys were to enroll and train a qualified Negro and then fly him in whatever vehicle is available, we could retell our whole space effort to the non-white world, which is most of it," he observed.[3] Webb told Murrow that NASA had all the astronauts it needed and that he was under pressure to appoint others, including women, to the astronaut corps. Later a NASA official stated that the space agency had resisted pressure to use the space program for what he termed political purposes and placed no premium on "special minority or political groups" in selecting who NASA considered the most qualified astronaut candidates.[4] NASA viewed advocacy of anyone other than white men as an issue for the public relations office or politicians on Capitol Hill.

Jerrie Cobb still had no clear idea what Webb had appointed her to do as a special consultant to the office of NASA's administrator. There had been no response to her report on women in space and no request for her services.[5] It was not even clear to whom she should report or how much she would be paid—if anything.* Rejecting Randy Lovelace's advice to keep silent, Cobb continued to speak about the women-in-space project before aviation groups, universities, and chambers of commerce across the country. She never, however, had anything official to do for NASA.[6] In fact, Cobb's public

---

*A handwritten report issued by the NASA Records Section on March 15, 1963, indicates that Jerrie Cobb was never paid anything for her services as a consultant at NASA. William V. Vitale, chief of NASA's Management Services Branch, wrote "Never paid anything!" on a single sheet that he placed in Cobb's employment folder. Later, another staffer at the NASA Records Section wrote "Very slim pickin's" in reference to standard employment forms that were present, or not present, in Cobb's NASA files.

comments on women astronauts were beginning to exasperate the space agency and Webb questioned if she should continue as a consultant. If NASA's plan had been to co-opt Cobb by making her a space agency insider, the objective failed. Cobb took every opportunity not only to speak about her own testing but about the other women of the Mercury 13 who also were waiting to become astronauts. In December 1961, when a Los Angeles aviation journalist organized a symposium on women in space, Cobb was invited to speak as a NASA representative. After space agency officials heard about the invitation, they cautioned the conference organizer that Cobb might not be the best choice, suggesting that Cobb was about to accept a commercial position which "may cause us some problems."[7] A week later, James Webb applied more direct pressure, telling Cobb that "since we have not found a productive relationship that could fit into our program, I am wondering if there is any advantage to continue the consulting arrangements." NASA would continue to look to military jet pilots as astronauts, he said, and no one could think of anything for her to do.[8] A week before Christmas, NASA again wrote the space symposium organizer and announced that Cobb's position at NASA was being terminated.[9] Cobb later maintained that while Webb had warned her that the consultant position was not working out, she received no official termination notice.[10]

Before traveling to Los Angeles for the symposium, Cobb updated the Mercury 13. She spoke of recent meetings in Washington with NASA and her search for answers to the Pensacola cancellation. "When I press for answers, they give me reasons," she wrote, "but none of them legitimate." She urged the women to continue avoiding publicity but indicated that "the time may be coming soon when I'll ask each one of you to make a small roar."[11] Cobb was planning to roar herself, and the women's space symposium presented the right opportunity. "I intend to keep hammering and trust that you all are still behind me," she told the Mercury 13.[12] At the symposium, Cobb was introduced as the nation's first woman astronaut and a NASA consultant. Cobb announced to the crowd that twelve other women also had recently passed tests at the Lovelace Foundation and illustrated her talk with *Life* magazine photographs depicting her tests at Albuquerque, Los Alamos, Oklahoma City, and even Pensacola. "This is just the beginning of the space race," Cobb declared, "an age we should all be proud to be living in." With her voice rising to match her rhetoric, Cobb unleashed a final note that was part benediction and part rallying cry. "The race for space will not be a short one—nor an easy one—but it is one in which we must *all* participate. Let us go forward, then—there *IS* space for women!"[13]

For the organizer of the conference, the response to Cobb's appearance was more than she had bargained for. Initially Jackie Cochran had been invited to deliver the keynote address. Cochran declined, citing scheduling problems. When Cochran learned that Jerrie Cobb had accepted instead, she called back and indicated that her schedule suddenly was clear. Deciding she could not rescind an invitation that had already been accepted, the organizer explained to Cochran that she would not replace Cobb. Cochran refused to accept the decision and invited the journalist to her ranch. Jackie Cochran did "some mighty arm-twisting," the conference organizer recalled—but her appeals did not work. So relentless was Cochran's argument over dinner that Floyd Odlum finally stepped in and admonished his wife to "lay off."[14]

While waiting for John Glenn to blast off the launch pad and become the first American to orbit the earth, Jackie Cochran and Jerrie Cobb finally agreed to meet. Each woman had come to Cape Canaveral to witness the historic moment and to absorb some of the limelight themselves. Jackie Cochran, using her influence, had a prime seat in Mercury Control.[15] Glenn's launch had come at the end of a long series of starts and stops and interminable delays. Over several months NASA had scheduled Glenn's liftoff ten times, and each time the launch date had been moved or the flight scrubbed. Finally, on February 20, 1962, at 9:47 A.M., Glenn roared off the launch pad. Glenn's flight aboard *Friendship 7* lasted just under five hours and propelled him into history. President Kennedy, speaking to the nation, declared that spaceflight represented humankind's most exciting adventure. "This is the new ocean," he said, "and I believe the United States must sail on it and be in a position second to none."[16]

During the final hold and long wait for Glenn's launch, Jackie Cochran and Jerrie Cobb sat down to dinner at a Cocoa Beach restaurant not far from Cape Canaveral. Cochran was surprised when Cobb asked to bring along Jane Rieker, whom Floyd Odlum later curtly described as "some woman writer for *Life* magazine."[17] Although annoyed, Cochran disguised her irritation with courtesy. Bringing Rieker to the dinner was a smart move on Cobb's part. Rieker was Cobb's trusted ally: a good talker who was at ease in social situations. Ivy Coffey once remarked that Jerrie Cobb seemed to attract people who wanted to help her. Dr. Shurley similarly observed that Cobb seemed dependent on Rieker, an aggressive, fearless, highly verbal woman.[18] Over the last two years, as much as Cobb had learned about being a public person, she still needed someone who could serve as her advocate. One thing was clear: Rieker did not come to the dinner to make Jacqueline Cochran's acquaintance. She was there to stand guard.

The dinner was a frustrating experience—at least for Cochran. She called her husband that evening and told him she was exasperated. Again and again, she said, Cobb asked her how she viewed a women-in-space program. After each question Cochran replied with the same answer, but the answer did not seem to sink in and did not appear to be what Cobb wanted to hear. So Cobb asked again and Cochran responded again. They got nowhere. A month later, still feeling stymied that she had not gotten her points across, Cochran wrote Cobb and spelled out her thoughts in writing. "Each time you asked me at Cocoa Beach how I felt about such a woman's program I tied to my expressions of approval some such words as 'if soundly organized.' So what I mean by this must be what is in doubt in your mind."[19] Cochran clearly laid out her ideas. From the perspective of national defense, there were plenty of qualified male candidates for spaceflight and they should go first, she argued. A larger number of women should be tested in order to reach broader conclusions about females' capability. Any hastily conceived program to put an American woman in space would be regarded as unnecessary drama. "Women for one reason or another have always come into each phase of aviation a little behind their brothers. They should, I believe, accept this delay and not get into the hair of the public authorities about it. Their time will come and pushing too hard just now could possibly retard rather than speed that date. It's better to be sound than quick."[20] Cochran advocated patience, deference, and acceptance of women's secondary status. Cobb could not have viewed things more differently.

"I am not content to sit back and listen to their silly excuses," Cobb wrote to Cochran, shedding her reserve.[21] If anyone seriously doubted that women should be astronauts, let them prove it by training and testing women to see how they measure up, she fired back. "The qualification rules have been laid down for astronauts and although NASA says they have nothing against women, it just so happens that the requirements are such that no woman can meet them." The requirements seemed inherently and even intentionally unfair to her: all astronauts must be test pilots, but no women can be test pilots. To Cobb, the restriction was a dizzying circle of irrationality. In an attempt to clear through such senseless reasoning, Cobb pointed to the exception NASA had made for John Glenn, who did not have the required college degree yet who had been allowed to become an astronaut based on his "equivalent experience." If NASA and the U.S. military would not permit the women to get jet time, Cobb believed that women should be acknowledged for their equivalent flying experience. Flight hours alone should count for something, Cobb contended, since a pilot always encountered emergencies, unforeseen problems that tested her ability to respond effectively to challenges. Jan Dietrich, Irene Leverton, B Steadman, and Jerrie Cobb had more flight hours than any of the seven Mercury astronauts.* Scott Carpenter, for example, had spent only 2,900 hours in the air and 400 hours in a jet. Flying thousands of hours over millions of miles in many different kinds of aircraft made these women just as quick, cautious, and able as a jet pilot. Realizing that the tenor of her debate was intensifying, Cobb tried to find some common ground that she shared with Cochran. They both wanted to see women in space, Cobb finally acknowledged, but they disagreed totally on the way to get there. Cobb envisioned an immediate women's training program—a concentrated effort for preparing an American to be the first woman in space. Cochran wanted women to stand in line. Having placed their alternate visions on the table, each woman set out to rally the Mercury 13 and then take their case to Capitol Hill.

Cochran went first. She made copies of her Cocoa Beach letter to Cobb and sent it out widely, starting with the Mercury 13. Cochran wanted each of the women candidates to know that she did not support a crash program for women astronauts and she was concerned that Cobb had misrepresented her point of view. Cochran's letter made little impression on most of the women. Many barely remembered reading it since some, like Sarah Gorelick, were rushing to find new jobs after the Pensacola cancellation.

---

*Jan Dietrich had 8,000 hours, Leverton 9,000, Steadman 8,000, and Cobb 10,000 hours. Scott Carpenter had 2,900 hours, Gordon Cooper 2,600, John Glenn 5,100, Virgil Grissom 3,400, Walter Schirra 3,200, Alan Shepard 3,700, and Deke Slayton 3,600. All the men had jet time that ranged from 400 hours for Carpenter to 2,500 hours for Grissom.

Gorelick decided to keep her calendar flexible in case testing was rescheduled, so she went to work as a bookkeeper in her father's retail store.[22] Although many of them ignored the letter, Cochran did win some support among the twelve women. Gene Nora Stumbough responded to Cochran, saying she agreed with her thoughts "100%."* While wanting to see women in space eventually, Stumbough wrote, "I am only afraid that by nagging those who make the decisions, we are hurting ourselves." There is no need to train women right now, she said.[23] In addition to the Mercury 13, Cochran also mailed the Cocoa Beach letter to Randy Lovelace, James Webb, other NASA officials, politicians, and Air Force Chief of Staff General Curtis LeMay. As was her practice, Cochran made sure her opinions were stated clearly and her positions well known, especially to men in power. As was also her practice, she kept meticulous files from her national newspaper clipping service of Jerrie Cobb's public appearances and press briefings. Cochran had only to ask her secretary for the file to know what Cobb was saying.

Jerrie Cobb, meanwhile, got in touch with Janey Hart, whose extensive Washington connections could be invaluable. What Cobb did not know yet was that Janey Hart was a formidable fighter in her own right. More outspoken than her husband and as savvy about strategy as any politician in Washington, Hart was the daughter of Detroit millionaire Walter Briggs, founder of Briggs Manufacturing, the largest auto body plant in the world. Hart grew up in a world of governesses, ocean liners to Europe, and private boarding school. As a teenager, Hart's world view changed when one of her teachers, a Catholic nun named Alma Miller, began arguing New Deal politics with her and encouraged Hart to keep a journal recording her questions about government and social welfare. Hart's political activism was awakened. She sided with labor unions, argued against segregation, and spoke out against restrictions the Catholic Church placed on women. By the late 1940s she was directly involved in politics, organizing Democratic grassroots campaigns and picking up the phone to call any relevant government official when she had an opinion to express.† The only time Hart avoided confronta-

---

*In reflecting on the letter forty years later, Stumbough Jessen said she had some regrets in the way she articulated her thoughts. She sensed that they could have been perceived as not supporting Cobb and Hart and the cause they believed in.

†Janey Hart was always involved in her husband's political campaigns. When Philip Hart started his first run for the Senate in 1957 and needed a way to quickly travel to small towns and state fairs, Janey Hart earned a helicopter pilot's license. "Number 25 in the

tion was when she was so outraged she knew she could not—as she put it—mind her manners. Hart admitted, however, that she "blew her stack" after the Pensacola tests were scrapped, clearly seeing the cancellation as an act of discrimination against women. Hart also recognized, as few of the other Mercury 13 did, that NASA's dismissal of the women was part of a larger system of social bias that restricted women's opportunity in nearly every aspect of American life. She believed that NASA's discrimination was part of the same system that, for example, restricted the number of women who could enroll in law schools, that did not provide young girls with athletic teams in public high schools, that required a husband's or father's signature in order for a woman to rent a car. In Hart's opinion, NASA's cancellation of the Pensacola tests was just one more example of intractable sexism that existed in American schools, business, political structures, even churches. Hart decided she could no longer honor Randy Lovelace's request to remain quiet about the astronaut tests.[24]

First Janey Hart dialed Liz Carpenter in the Vice President's office. Although the two women had never met, they had operated in the same national Democratic circles for years and shared a respect and fondness for Lady Bird Johnson.* Intuitively, Janey Hart knew she could trust Carpenter. Hart had seen Carpenter plenty of times at meetings and heard her on the radio as press representative for Lady Bird and now for Vice President Lyndon Johnson. She knew they shared a similar political philosophy and that Carpenter believed strongly in extending opportunities to women. Hart held two ideals particularly dear: be true to your principles, and always have fun. Carpenter seemed to do both. Once when Liz Carpenter was being vetted for a position for an incoming administration, she was asked if anyone in her family or circle of friends had a background that could possibly embarrass

free world," she later recalled. Phil Hart was not Janey Hart's only political passenger. In 1960, when John Kennedy sent his mother to campaign in towns across Michigan, Rose Kennedy flew from Mackinac to Battle Creek with Janey Hart in the cockpit.

*Janey Hart believed Lady Bird Johnson understood her irritation with the roles Senate wives were supposed to play in Washington, D.C. As the wife of the Vice President, Lady Bird Johnson was required by political custom to chair a committee of Senate wives who participated in volunteer activities such as rolling bandages while dressed in Red Cross uniforms. Hart refused to participate, finding the ritual demeaning, even archaic. "Turn-of-the-century stuff," she later called it. Hart told Lady Bird Johnson, "You make the bandages and I'll fly them anywhere you want," adding, "I would rather roll a bandage up Connecticut Avenue at the end of my nose than go to a thing like that." Lady Bird Johnson apparently was amused and let Hart off the hook.

the President. Carpenter's answer was, "Yes, thousands." Hart loved that story, which was widely circulated among Democrats.[25]

Carpenter listened as Hart laid out the story: "phase one" Lovelace Foundation testing in Albuquerque, "phase two" isolation testing in Oklahoma City, and "phase three" spaceflight simulation testing abruptly canceled for lack of a NASA "requirement." Carpenter laughed when Hart pointed out that women seemed to have at least as much sense as Enos, a NASA space monkey that had been launched. Why should outer space be for men only, Hart asked? Carpenter agreed to review a packet of materials assembled by Senator Philip Hart's staff outlining the argument for a women's astronaut program. She also agreed to speak directly with the Vice President and ask if he would be willing to meet with Hart and Jerrie Cobb to discuss their views.[26]

While waiting for an appointment with Lyndon Johnson, and with Carpenter calling NASA for its side of the story, Hart wrote letters to every member of both the U.S. House and Senate space committees and enclosed Cobb's recent speech from the space symposium. Hart called NASA's argument that women need experience as jet test pilots specious and a neat dodge. The reason women did not have test pilot experience, she wrote, was that they were barred from flying in the military and no civilian companies would hire them. Hart wasted no time with red herrings. She went right to what she saw as the source of the discrimination. NASA refused to sanction testing of women astronaut candidates because women astronauts posed a threat to men's fragile sense of masculinity. A space program for women could have been launched years earlier, she said, if some men could be convinced that a woman at the controls of a space capsule "would not destroy their virility."[27] Hart released her congressional letter to the press, and that was when the phone started ringing at the Harts' home. "What does your husband, the Senator, think about your petition to Congress?" a reporter asked. "I've never asked him," Hart replied.[28]

Hart's lobbying worked. Within days of her request for an interview, the Vice President agreed to meet and Hart wired Jerrie Cobb in Oklahoma asking her to leave immediately for Washington. Lyndon Johnson was an especially important ally to win. Not only had Johnson guided the legislation that had created NASA in 1958, he was also currently serving as head of the President's Space Council and John Kennedy's liaison with NASA. More important, the Vice President also understood that space travel promised greater benefits to the country than scientific research alone. Like Kennedy, he was keenly aware that spaceflight could inspire a nation. Johnson had been so enthused with John Glenn's orbit that he had flown to Grand Turk Island to personally accompany Glenn back to Cape Canaveral.[29] And Johnson wanted to be recognized publicly as the administration's link to space. When Glenn and his family later made their triumphant ride to the United States Capitol, Johnson was squeezed into the front seat of the astronaut's open convertible. People lining the streets on that rainy day in Washington focused almost as much on the Vice President as they did on Glenn. Lyndon Johnson's massive build and large gestures all but overwhelmed the car. He looked like a big, excited bloodhound—all ears and gaping mouth and insistent head.

Before the meeting with Jerrie Cobb and Janey Hart, Lyndon Johnson quickly scanned the background material Liz Carpenter had prepared. She urged the Vice President to give the women some encouragement and drafted a letter to NASA's James Webb for Johnson's signature. The letter, while not a ringing endorsement of women astronauts, asked Webb if any women met NASA qualifications for astronauts or if any woman had been disqualified simply because she was a woman. "I'm sure you agree that sex should not be a reason for disqualifying a candidate for space flight," Carpenter drafted in Johnson's letter. Along with the proposed letter, Carpenter also provided summaries of both sides of the argument. She wrote that Cobb and Hart believed that women should be used as astronauts in order to beat the Russians in sending the first woman into outer space and to demonstrate that space was not for men only. NASA, however, believed that spaceflight still was too risky and that once orbiting became a more routine practice, women would be considered. Carpenter also cited NASA's insistence on jet test pilot experience. Her recommendations to Johnson were clear: hear the women's petition, show them the letter to James Webb, and offer them some support. "I think you could get good press out of this if you can tell Mrs. Hart and Miss Cobb something affirmative," Carpenter wrote in her memo to the Vice President. "The story about women astronauts is getting a big play and I hate for them to come here and not go away with some encouragement."[30]

Indeed, the issue of women in space was getting publicity in the press, thanks to interviews Hart and Cobb had been conducting with reporters. The morning of Cobb and Hart's meeting with Lyndon Johnson, Congressman Ken Hechler of West Virginia called for women astronauts, posting in the *Congressional Record* a copy of Cobb's recent speech before the women's space symposium. Hechler prefaced Cobb's address with comments in his capacity as a member of the House Committee on Science and Astronautics. "I believe that we should give serious consideration to the inclusion of women among our future astronauts," he said.[31] Columnists across the country were also weighing in. A science writer for the *Dallas Times Herald* offered a not-so-subtle condemnation of the idea: "Let them vote. Let them wear pants. Let them shoot pool. But please, Mr. Vice President, don't let them get into space." The column presented a mock conversation of women astronauts at the controls of an orbiting space capsule—dialogue that was filled with stereotypes about women's lack of technical knowledge, fascination with interior decorating, and absentmindedness. "The little thingamabob has jiggled off the gizmo," the woman astronaut reported. The column was illustrated with a drawing of "Geranium 7"—adorned with tieback curtains and winding garlands of flowers. A caricature of LBJ looked in horror at the frilly capsule.[32]

Jerrie Cobb arrived in Washington the day before in order to meet with an old acquaintance from Oklahoma, Senator Robert Kerr, who now served as chairman of the Committee on Aeronautical and Space Sciences. Kerr threw a news blackout on their meeting in an effort to avoid any publicity about women astronauts coming from his office. Although others saw Kerr's blackout as a way for the senator to distance himself from the issue of women in space, Cobb believed the senator appeared willing to offer any help he could.[33] Cobb read him wrong, however. Over the coming months, Kerr never used his considerable power to persuade his colleagues on Capitol Hill or at NASA that Cobb and other women deserved a chance to be astronauts. As she frequently did, Cobb made the mistake of viewing courtesy as commitment. She met Janey Hart the next day equally hopeful that Lyndon Johnson would be able to help them.

Cobb and Hart had little time to get to know each other and quickly reviewed the points they hoped to make and the best strategy to use. They wanted to impress upon the Vice President the importance of female astronauts for the scientific data a woman's flight would produce and for the chance to beat the Russians. Hart believed Russia would launch a woman as soon as the fall. Cobb wanted to push for a concentrated women's training program that would begin immediately. One issue they did not discuss was what the other women pilots wanted. Almost a year after the Mercury 13 passed the astronaut physical tests at Albuquerque, their identities still had not been publicly revealed. Cobb knew all their names. So did Jackie Cochran, who continually asked Randy Lovelace for updates on the women's addresses. Even the women themselves did not know who the other successful candidates were. Cobb's updates were always addressed to "F.L.A.T.s" and never listed the women's individual names. They had not heard a word about the thirteen from Randy Lovelace since the Pensacola telegram. Most important, the women as a whole had never met to discuss their objectives or strategy. The group meeting that Lovelace had proposed had never happened, since the Navy had canceled the Pensacola tests before the women had a chance to gather in Florida. As a result, Cobb and Hart prepared for the meeting with Johnson without the insight, debate, and support of the larger group.[34] They were unaware, for example, that Gene Nora Stumbough was not an advocate for an immediate women's astronaut program. They also were unaware of some connections that might have been useful in making their points, such as Jerri Sloan's acquaintance with Lyndon Johnson. As they climbed the main staircase to the Senate reception room at the Capitol, Janey Hart and Jerrie Cobb hoped their words would be enough.

The Vice President agreed to meet with the women at 11 A.M. in his office across from the Senate chambers. A master of political real estate, Johnson maintained his Senate leadership office after he assumed the vice presidency. Journalists around Capitol Hill referred to the elaborate chambers of Office P-38 as the "Taj Mahal"; it was an impressive room with views of the Supreme Court, a large crystal chandelier, and ornate frescoes on the ceiling.[35] Johnson rushed to the meeting after attending a bill-signing ceremony at the White House with President Kennedy. The Vice President had just an hour to meet with the women, grab something to eat, and prepare to open the Senate at noon. As Johnson walked out to greet them in his reception area, Cobb and Hart gathered up their pocketbooks and extended their hands in greeting.

Cobb focused immediately on the scientific benefits that could be gained by sending a woman into space. She presented Johnson with the same points she had been making for nearly two years: women weighed less, ate less, consumed less oxygen than men. Therefore, women would need less booster power to propel them into space. Recent studies, she explained, proved that women showed an amazing ability to withstand isolation and inactivity. She reviewed the tests in the isolation tank that she had completed with Dr. Shurley in Oklahoma City and indicated that Wally Funk and Rhea Hurrle had performed equally well. New research, she continued, revealed that women could withstand more heat, noise, and vibration than men.[36] With such results, Cobb argued, how could the United States government discontinue testing of women astronaut candidates?[37]

Hart then added her points. Space should not be blocked off as an environment for men only, she said. It was an antiquated idea to suggest that women only wanted to stay home, tied to the kitchen. Women wanted to explore the universe and push themselves to the far reaches of their ability, just as men did. Besides, opening this door to women was part of a larger national effort toward equity and fairness for all Americans. As Johnson knew, President Kennedy himself had announced on the very day that John Glenn had orbited the earth that he was establishing a Commission on the Status of Women. In an executive order posted at all government agencies, the President made it clear that "women are entitled to equality of opportunity for employment in Government and in industry. But a mere statement supporting equality of opportunity must be implemented by affirmative steps to see that the doors are really open for training, selection, advancement and equal pay."[38] Eventually women would explore outer space, Hart argued. Why not begin at the earliest moment we can?[39]

Johnson leaned back in his chair. Above him on the ceiling were four allegorical frescoes depicting human ambition. All four figures were women, dressed in impressive robes and staring down at him.[40] Johnson folded his hands and leaned his large shoulders against the back of his leather chair. Jackie Cochran had turned him in favor of women fliers a long time ago, he began.[41] In fact, Cochran had once saved Johnson's political neck, an aide later recalled. In the final days of his 1948 Senate primary race, Johnson had collapsed in pain from a kidney stone, high fever, and dangerous infection. He could not afford to take time off for surgery—the days lost on the campaign trail would be politically fatal. When Jackie Cochran heard about his plight, she recalled that an esteemed urologist from Great Britain was visiting the Mayo Clinic. Cochran called one of Johnson's campaign managers and informed him she would appear at the hospital emergency entrance at 1:30 and fly Johnson to Minnesota in her Lockheed Electra. She did. Johnson was relieved of his pain without surgery and was back on the campaign trail in two weeks. Johnson did not forget personal favors like that.[42]

Many minority groups were asking for attention from NASA, the Vice President continued. They wanted to be astronauts, too. If the United States allowed women in space, then blacks, Mexicans, Chinese, and other minorities would want to fly too. Cobb sat, listening politely, looking prim in her tailored dress, with three strands of pearls around her neck. What's wrong with minorities serving as astronauts if they are qualified, she asked? Johnson did not answer. Cobb continued. If the Vice President were proposing that only citizens who were in the majority should be launched into space, then women should be considered. Women are certainly not in the minority, she thought, in terms of numbers, money, votes, and tax dollars. Leaning toward the women with a pained expression on his face, Lyndon Johnson looked directly at Cobb and Hart and gave them his final thought. As much as he would like to help the cause of women astronauts, it was really an issue for James Webb and those at NASA. It hurt him to have to say it because he was eager to help, but the question just was not up to him to address. Johnson called an end to the meeting and started talking on his private telephone.[43]

Janey Hart was angry. She knew Johnson was "putting on a performance that made it look as though it was painful to tell us."[44] Clearly Johnson was not going to lend a hand to their cause, even though a word from the Vice President to James Webb would make an enormous difference. What Hart did not understand was why.

Hart and Cobb left Johnson's chambers and met with a crowd of reporters outside in the Capitol hallway. Hart stood with her arms tightly folded across her chest, her pocketbook stuffed into the crook of her arm. Her goal at this point seemed to be to mind her manners and hold her anger in check. Cobb leaned near the wall, her face set rigidly in a practiced smile. "I'm hoping that something will come of these meetings," she politely said as reporters scribbled into their notebooks. Later, newspaper reports declared that two would-be "astronettes" had pleaded their case in Washington. The Vice President—using the current jargon from Cape Canaveral—had said the women were "A-OK" but the decision was not his to make.[45]

Cobb and Hart never saw the letter to James Webb that Liz Carpenter had drafted for the Vice President's signature.[46] Johnson decided not to show it to them because he had no intention of signing it. He did not want to ask James Webb to look into the question of women astronauts. Perhaps Johnson thought starting a woman's program would jeopardize the whole works, Carpenter later said.[47] Taking out his pen, Johnson drew Carpenter's draft across the large desk and scribbled forcefully across the bottom of the page. In his distinctive hand, Johnson announced the verdict that Hart, Cobb, and the press never knew: "Lets Stop This Now!"[48]

Congressional Hearings: Sound and Proper

AFTER NEWS BROKE ABOUT COBB AND HART'S MEETING, LETTERS began piling up in the Vice President's office and were shuttled off to young interns for reply. Most of the letters were from women urging Johnson to support Jerrie Cobb and Janey Hart in their campaign. "I think the intelligence, patriotism, initiative and creative ability of women is our most wasted resource in this country," one woman wrote.[1] Other women, inspired by the idea of women in space, volunteered themselves for astronaut service. "I am an American Negro Woman. I am in my 40s. I am in good health. I think there is a job I can do in the Program. Please give me a chance to help."[2] Of course, not everyone agreed. One letter, signed simply "a Bachelor," complained that it was getting so a man could not go anywhere without having a woman hanging around.[3] Jerrie Cobb also received letters, perhaps none as supportive as the one from the director of research at the U.S. Naval School of Aviation Medicine in Pensacola. Like many doctors both at the Lovelace Foundation and in Pensacola, he wanted the women's astronaut program to go forward and was disappointed with the abrupt cancellation of further testing. He wrote Cobb congratulating her on lobbying Lyndon Johnson and then added that her efforts represented "the biggest crusade since you gals put through Woman's Suffrage."[4]

Unaware as they were of Johnson's directive to stop any women's astronaut program from moving ahead, Cobb and Hart did recognize that their appeal to the Vice President had not been met with enthusiasm. There was still one more avenue left to pursue—perhaps the only one—that might require NASA to go forward with the Pensacola tests. In March, Hart wrote members of the House and Senate space committees, urging them to consider calling a subcommittee hearing to determine if women were being discriminated against in the United States' space program.[5] Since Congress appropriated funding for NASA, James Webb would take seriously any findings from Capitol Hill. Hart's appeal to the Senate was unsuccessful. Oklahoma Senator Robert Kerr, the powerful chair of the Senate Committee on Aeronautical and Space Sciences, made no offer to hold a committee hearing. While Kerr was always willing to bask in the reflected limelight of Cobb's aviation accomplishments and while he proved exceptionally successful in bringing pork barrel space projects to Oklahoma, he would not lend a hand. He also was unwilling to privately urge his good friend Lyndon Johnson to reconsider his position. The Senate was not going to fight for women astronauts. The House of Representatives, however, showed more interest in Hart's letter. A chance meeting between Jerrie Cobb and the House space committee chairman generated additional support for a hearing.[6] By the middle of June it was official: the House Committee on Science and Astronautics would investigate alleged government discrimination against women in the nation's space program. Preliminary committee research would begin immediately, and subcommittee hearings would be held in July.[7]

Democratic New York Congressman Victor Anfuso accepted the chairmanship of the eleven-member special subcommittee, which included the full committee's only two women members.* While pledging to keep an open mind, Anfuso believed that Cobb and Hart might have a point. Some changes in the current law may be necessary, he admitted. Anfuso knew that only

---

*The special subcommittee's two women members were Jessica McCullough Weis, Republican of New York, and Corinne Boyd Riley, Democrat of South Carolina. Riley, a "widow's mandate" member of Congress, was elected to fill out the term of her deceased husband. She entered Congress only three months before the astronaut hearings, inexperienced and with scant knowledge of space issues. She contributed little to the hearings and spoke only when coaxed into the discussion by special subcommittee chairman Anfuso. Weis served in Congress from 1959 to 1963. She was not a candidate for reelection in 1962 due to ill health. She died the following year.

men were allowed to become military jet test pilots and that no woman could meet that NASA requirement.[8] Initial plans for the hearings were ambitious. Anfuso wanted to call many witnesses, including the Mercury 13, NASA officials, Dr. Randy Lovelace, Jacqueline Cochran, women scientists, perhaps former First Ladies Eleanor Roosevelt and Bess Truman. The hearings in Washington would take place over three days, with subsequent hearings slated for New York, the Midwest, and California.[9]

Jackie Cochran, reluctant to talk with Jerrie Cobb after what she viewed as their conversation fiasco in Cocoa Beach, called Janey Hart and invited her to lunch at Cochran's spacious apartment in Manhattan. Hart knew her liberal political views were apt to clash with Cochran's conservative perspective. Six years earlier, Cochran had run unsuccessfully for Congress as a Republican in her California district. She knew a great deal about Capitol Hill and certainly regarded Phil Hart as one of the most liberal voices in the Senate.

But Hart was practiced in presenting her views to unreceptive audiences. She also understood that it was more important to persuade Cochran than to preach. She would listen to what Cochran had to say and then offer her opinion. Many times she had campaigned for her husband in conservative areas around Detroit. She was used to Senator Hart's press secretary standing in the back of the room, arms folded and wincing, if she veered too far to the left. At first sight of a cringe, Hart would draw herself in and temper her remarks.[10] No press secretary would be giving Hart signals at Jackie Cochran's dining room table, but Hart thought she could monitor the temperature in the room.

Cochran's New York foyer, like her home in Indio, projected an immediate sense of power and accomplishment. Cochran's many trophies were displayed in the central hall: world records, air races, presidential citations. Painted on the floor of the foyer was an enormous aviator's compass. Hart headed south into the dining room, stepping over the marking in the exact center of the compass—"Starting point," it read. During the next several hours, as much as Hart tried, she could not persuade Cochran that the Mercury 13 should be allowed to continue their testing in order to be considered as serious astronaut candidates. Cochran believed everyone should start over and organize a larger testing program for women. With only thirteen women coming out of the Lovelace exams, she feared too many would marry or have children and the program would be left with inadequate numbers. She also believed that men came first and women should not interfere with the serious objectives of the space program. Hart listened as respectfully as she could but could not believe what she was hearing. What shocked Hart the most was that Jackie Cochran was advancing the exact arguments that she had fought so hard to defeat while organizing the WASPs. Twenty years earlier, Cochran had confronted hostile generals, a reluctant government, and threatened airmen, and convinced them that training women to fly Army Air Corps planes was not a waste of money. Women had talent and wanted to serve their country, Cochran had argued. Now she was using the opposite reasoning to stall the women's astronaut program. Hart was baffled as to why Cochran had poured money into women's astronaut testing a year earlier and now wanted to stop it from advancing. "Something bothered and bothered her," Hart later said. Maybe she just wanted more trophies for the foyer, more headlines and accolades. If Jackie Cochran could not become the first woman in space, perhaps she did not want any other women to have the chance.*[11]

As much as Hart did not like what she heard at lunch, she had to admit that Jackie Cochran had been straightforward with her. Shortly after their conversation, Cochran sent Hart a draft of the testimony she expected to deliver at the upcoming subcommittee hearings, now just a month away.[12] Cochran sent her draft to others as well. She circulated it to James Webb and other officials at NASA, the Air Force chief of staff, Navy admirals at the Pentagon, and Randy Lovelace. She also sent a draft to Gene Nora Stumbough, since Cochran felt she had an ally in the young woman pilot. Cochran suggested that if Stumbough could not be present at the hearings, she should forward any comments directly to her. Cochran wanted to keep Cobb out of her conversations as much as possible and she wanted to take advantage of divergent views among the Mercury 13.[13]

Jerrie Cobb did not realize there were divergent views among the

---

*Others agreed with Hart. Syndicated columnist Phyllis Battelle used an article to question why Cochran did not support an immediate women's astronaut program. Battelle asked, "Could it be she is protecting her own fantastic number of speed, distance, and altitude records?" Voicing her opinion on Cochran's concern that women astronauts might become pregnant, Battelle claimed, "I'd bet my last fond memory of the Bobbsey Twins that the women involved in a smartly-organized space effort would take every precaution to prevent problems of health and heart from interfering with their roles in scientific history."

Mercury 13. She became their spokeswoman simply because she stepped to the microphone first. Cobb often said that she began speaking for the group by default. In fact, Jerrie Cobb, Janey Hart, the Dietrich twins, and Sarah Gorelick were the only women whose connection to the women astronaut program was publicly known.* Most of the other women still complied with their pledges of secrecy and kept their identities hidden, not wanting to do anything that could jeopardize their chances. In addition to speaking for the group, Cobb also clearly saw herself as the leader. She sent the other women updates on whom she had petitioned, circulated press clippings, and spoke about their testing in the many speeches she made across the country. Grateful for the liaison work Cobb was doing and indebted to her for vigorously pushing for Pensacola exams, most of the women accepted Cobb as their representative. Yet as the hearing neared, some of the women, such as her good friend Jerri Sloan, wished Cobb would involve the other women more. In situations where Cobb needed to speak as a public person, she still kept a trusted woman ally such as the vocal Jane Rieker in tow. But when it came to making decisions about the women-in-space program, Cobb made them alone. Cobb was so used to flying solo, to acting independently, Sloan later said, that it often never occurred to her to bring others into the decisions.[14] In fact, there was another reason that Cobb did not assemble the women and ask for their opinions: she thought she knew more than they did. Since the other women were not aware of the facts or inner workings of the program, Cobb later said, they were not in a position to know what actions should be taken. When it came to planning strategy and deciding whom to prod and how hard to push, Cobb made the calls.

Jerrie Cobb was optimistic as she flew her Aero Commander from Oklahoma City to Washington to testify before the subcommittee. Although she had plenty of room for other passengers in the plane, Cobb did not take along any of the thirteen, such as Jerri Sloan in nearby Dallas or Sarah

---

*Jackie Cochran played a role in revealing the identities of both the Dietrichs and Gorelick. Even though it annoyed Cochran that Cobb generated publicity, Cochran wrote the article for *Parade* magazine that identified the Dietrich twins as Lovelace candidates. Cochran also revealed Gorelick's identity. When Cochran passed through Gorelick's hometown of Kansas City on a speaking engagement, she told reporters about the testing and Gorelick's role. Gorelick later admitted she had been "flabbergasted" at Cochran's disclosure. "We were sworn to secrecy about not telling anyone about the tests," Gorelick recalled, "and here was Jackie breaking the rule."

Gorelick in Kansas City. Instead, she asked *Life* magazine writer Jane Rieker and Cathryn Walters, Dr. Shurley's research assistant in the isolation experiment, to join her. Walters had become a supporter of Cobb and was ready to testify, if called upon, concerning Cobb's record run in the tank. The subcommittee, which was responsible for selecting witnesses, had not invited Walters.[15] The number of witnesses asked to testify had been dramatically reduced over the month of preparation. Anfuso also scaled back the number of hearings themselves, canceling the testimony in New York, the Midwest, and California, preserving only the three days of hearings in Washington. Only Cobb and Janey Hart would speak for the Mercury 13. For NASA, the witnesses were George Low, director of spacecraft and flight missions in the Office of Manned Space Flight, and astronaut John Glenn, who had recently completed his orbit around the earth. Cobb learned that the reason the hearings had been cut back was that the chairman of the House Committee on Science and Astronautics, George Miller, had overruled Anfuso and instructed the subcommittee chairman that only two representatives could be heard from each side. Jackie Cochran then used her political influence to persuade Miller, a fellow Californian, to include her in the hearings. With Cochran added to speak with Cobb and Hart, the subcommittee then permitted NASA to add astronaut Scott Carpenter as the third representative for their side.[16]

No other witnesses were officially invited, including Dr. Randy Lovelace or anyone from the Pentagon or the School of Aviation Medicine in Pensacola. Since he was the architect of the women's tests and NASA's chairman of the Special Committee on Life Sciences, Lovelace's absence was particularly conspicuous. No one knew if Lovelace had simply been overlooked or if he had decided that public advocacy of a women's program might hurt his chances for continued involvement with NASA projects.[17] He certainly was the one person who could have described the original intentions of the testing, its genesis with Don Flickinger, and the scientific contributions women astronauts could make to the space program. Jackie Cochran contacted Lovelace for his opinion prior to the hearings and he responded in an extensive memo. Lovelace told her that the only way scientists would find out how women might perform in space was to continue ground experiments such as the ones slated for Pensacola. He predicted that it would take as long as five years before meaningful scientific results on women could be achieved, two years longer than the men's testing and training. Lovelace wrote that he and Flickinger had started the program because they thought it would fill a need for a future women-in-space project. They did not expect a woman would sit in a space capsule immediately.[18] The night before the hearings began, Cochran checked in again by telephone with Lovelace. His opinion was unchanged: a woman should not immediately be injected into training at Cape Canaveral but testing the Mercury 13 should continue.

ON TUESDAY MORNING, JULY 17, JERRIE COBB WALKED INTO THE congressional committee hearing room, realizing that the next three days would determine her future. If there was any hope left of sending an American woman into space, that hope lay here, she thought.[19] It had been almost three years since she had met Don Flickinger and Randy Lovelace on the beach in Miami and heard of their interest in testing a woman pilot for astronaut viability. It had been nearly two years since she had surprised doctors with her physical and psychological stamina. Even though she met resistance from NASA, Senator Kerr, and Vice President Lyndon Johnson, Cobb believed that over the next three days she would be able to make her case.

As people filed into the hearing room, Cobb greeted Janey Hart and introduced her to Rieker and Walters. Hart worried that since she and Cobb had not compared drafts, their testimony might overlap. There was no time for edits, however. Anfuso ushered his subcommittee into their seats and the crowded audience and members of the press took their places in rows behind the witness table. Cobb would go first. She had carefully marked her draft with slashes between sentences to indicate where she should pause, what words to emphasize, and noted in the margins when she would smile. All the phrases describing the testing that had drawn laughter from previous audiences were inserted into the text. Cobb had practiced the five-minute speech many times and made notes about the attitude she should convey. "Never apologize, no timidity, in control," her small, precise printing reminded her.[20] For the reticent woman who twelve years earlier had walked out of her college speech class in frustration, testifying before a subcommittee of the U.S. Congress seemed almost as improbable as standing on the moon.

Precisely at ten o'clock Congressman Anfuso called the subcommittee to order. Hart had expressed concern to Cochran when Anfuso was appointed chair of the hearings. She knew from conversations on Capitol Hill that Anfuso had decided not to seek reelection and was worried that he might approach the hearings with a lack of interest, passively biding his time during the final months of his term.* "Ladies and gentlemen," he began. "We meet this morning to consider the very important problems of determining to the satisfaction of the committee what are the basic qualifications required for the selection and training of astronauts."[21] What Anfuso said next made Hart feel better. He seemed to understand that the women wanted the right to contribute their talents for a national cause. "We are particularly concerned that the talents required should not be prejudged or prequalified by the fact that these talents happen to be possessed by men and women. Rather, we are deeply concerned that all human resources be utilized," he said.[22] Anfuso laid out the day's agenda: following Jerrie Cobb's testimony, Mrs. Hart would speak. There would be questions, and later that morning, Miss Cochran—who had not yet arrived—would deliver her testimony.

Cobb leaned forward in her leather chair facing the committee and, in her slow, careful Oklahoma cadence, began her appeal. "We appreciate the vision and interest you are showing in recognizing the need for looking into the utilization of women in the U.S. space program on a serious and sound basis," she said. Cobb's last four words were especially well chosen. In every letter, every conversation, every objection, Cochran had used the same words to criticize Cobb's desire for an immediate program: it was not "sound" and it was not "proper." If Cobb could co-opt Cochran's words, use some phrases first and address the concerns they suggested, perhaps she could win the debate. The testing program in Albuquerque used no taxpayers' money, she continued, and described how she and the Mercury 13 had become involved. Why are those women pilots not present? she was asked. Her answer startled the audience: the women who successfully passed the tests did not even know one another. "They have never met as a group, and no one of the twelve women knows who all the other eleven are." Since no funds were available to bring them to Washington, the women could not come, she said.[23] Cobb's statement was almost true. Cobb knew the names of all the women and Hart did, too, after Jackie Cochran supplied her with a list.[24] Many of the thirteen knew some of the other women who had passed the

---

*Anfuso was elected a judge to the New York Supreme Court in 1962.

tests, but they did not know everyone. And while congressional funds to cover their travel to Washington would have made the trip possible for many of the women, some would have been happy to pay their own way in order to testify.[25] The problem was—they had not been invited.

In lieu of their individual testimonies, Cobb provided brief biographical portraits of each woman: Jan and Marion Dietrich from California; Rhea Hurrle Allison, now married and an executive pilot in Houston; Irene Leverton, second in total flight hours only to Cobb; B Steadman, owner and operator of her own aviation service; Jean Hixson, former WASP and Air Force Reserve captain who taught school in Ohio; Gene Nora Stumbough, a former university flight instructor; Jerri Sloan, officer of her own Dallas air services firm; Myrtle Cagle, civilian flight instructor at a Georgia Air Force Base; Sarah Gorelick, former engineer from Kansas City; and Wally Funk, only twenty-three years old with already 3,000 hours of flying. Reporters, hearing the women's names for the first time, tried frantically to scribble down each name. Anfuso allowed Cobb to spell each name and hometown.[26] Some reporters dashed off to nearby telephones to call their city editors. One reporter from Los Angeles called his newspaper office and asked for help in locating Jan Dietrich, Irene Leverton, and Wally Funk, who all lived in the LA area.*[27] Now that the women's names had been released, everyone wanted to know more about them. One of the congressmen from California even wondered aloud why Irene Leverton was not present for the hearings. Leverton lived in his district, and he wanted to meet her.[28]

Cobb continued with what she considered her most persuasive evidence—scientific research that indicated women were uniquely suited to spaceflight. She reeled off the arguments she had been making for three years about women's lower body weight, resistance to radiation, and ability to withstand isolation, heat, cold, noise, and pain. Who could not pay attention to the results of these studies? she thought. She looked at the eleven members of the committee and reminded them that throughout history women always had been eager to take part in humankind's boldest adventures. Women had been on the *Mayflower* and the first wagon trains, she

*Gene Nora Stumbough was also contacted by editors in Boise, Idaho. "I haven't heard a thing about it since [Pensacola]," Stumbough replied. "I believe it is a dead issue. I also think there won't be any women in the space program in, well, say the near future. All I lost was my job and 20 pounds."

said. Now women wanted to use their skills and courage for the next great expedition.

In writing her testimony weeks before, Cobb had shied away from lofty language that might have heightened the drama of the moment or fueled her remarks with indignation. Just like the flashy designer clothes that her sister encouraged her to wear for record-setting flights, inflated prose seemed phony and ill-fitting to Cobb, and she refused to wrap herself in it. Her final words to the subcommittee were remarkable in their plainness. To the audience who listened carefully, however, what Cobb said resounded with quiet dignity: "We seek only a place in our Nation's space future without discrimination. We ask as citizens of this Nation to be allowed to participate with seriousness and sincerity in the making of history now. . . . We offer you thirteen women pilot volunteers."[29]

The eloquence of Cobb's appeal was immediately undercut by Chairman Anfuso, who made a joke about women's reproductive capacity: "I think we can safely say at this time that the whole purpose of space exploration is to some day colonize these other planets and I don't see how we can do that without women." He smiled as the audience erupted in laughter. Anfuso quickly turned to Janey Hart and introduced her as the wife of a distinguished senator and the mother of eight children. Hart sized up the laughter and turned it to her advantage. "I couldn't help but notice that you call upon me immediately after you referred to colonizing space," she retorted, conscious that nearly everyone commented on her large family.[30] Hart felt she could afford the self-deprecating joke. She knew that the words she was about to deliver would underscore where she stood. More than anyone else in the room, Hart realized that the hearings were not just about thirteen women astronaut candidates. They were about every woman's equality.

The moment was Hart's first and last public testimony before the United States government and she wasted not a second. "It is inconceivable to me," she began, "that the world of outer space should be restricted to men only, like some sort of stag club. I am not arguing that women be admitted to space merely so that they won't feel discriminated against. I am arguing that they be admitted because they have a very real contribution to make."[31] Hart drew her arms up on the table, as congressmen looked curiously at the oversized pilot's watch Hart always wore on her left wrist.[32] Saying that women should not be allowed to go forward as astronauts was an idea as antiquated and wasteful as the assumption one hundred years ago that women should not be nurses, she said. A century ago, during the Civil War, wounded soldiers had covered battlefields across the country. Nursing care had been limited. Not enough male nurses could be found to tend to the dying. Women who wanted to help in the hospitals and nursing stations frequently were turned away by the men who staffed them. Women would faint at the sight of blood; they could not be trusted to keep medication straight. It was improper for young women to care for men who were strangers to them. Only middle-aged and ugly women were allowed to serve—presumably because ugly women had more strength of character, Hart continued, unable to resist a sarcastic jab. In rejecting the contributions of women, the country paid a terrible price. "I wonder if anyone has ever reflected on the great waste of talent," Hart declared, "resulting from the belated recognition of women's ability to heal." Knowing that Jackie Cochran would soon make her claim that women were not needed as astronauts because enough men already stood in line, Hart was emphatic. "It seems to me," she said, "a basic error in American thought that the only time women are allowed to make a full contribution to a better nation is when there is a manpower shortage." As a result, women become discouraged from developing their talents since it seems they will never be called upon to use them. "If girls elect to be homemakers, excellent—provided the choice is not dictated by discrimination in all other careers. . . . Let's face it," she declared, "For many women the PTA just is not enough."[33] Reaching the end of her statement, Hart offered one concrete suggestion: the testing at Pensacola should be continued. Even if women would not be launched into space immediately, the scientific data that could be gathered from testing healthy women would be invaluable, she said. Hart underscored her final point. "I don't want to downgrade the feminine role of wife, mother and homemaker," she said. "But I don't think, either, that it is unwomanly to be intelligent, to be courageous, to be energetic, to be anxious to contribute to human knowledge."[34]

Jerrie Cobb's simplicity and Janey Hart's fire combined to make a compelling case for women astronauts. Congressman Anfuso opened the question-and-answer session by thanking Hart for her "interesting" statement.[35] Quickly the questions turned to the stumbling block of military jet test experience. Anfuso asked Cobb if she thought jet test pilot experience was essential to being an astronaut. No, Cobb replied. The job in space was to pilot a spacecraft, not test a jet. Equivalent flying experience should count, Cobb argued. If NASA insisted on jet experience, then women should have the chance to train in them, or at least in jet simulators. All the time, Cobb kept thinking about the taxes millions of American women paid for military jets and training. Why shouldn't thirteen women have access to that equipment? she wondered.

Jessica Weis, Republican congresswoman from New York, spoke up. Who was responsible for canceling the Pensacola tests, and what were their objections? she asked. Weis saw no reason why the tests could not go forward, at least for the scientific data they would provide. Cobb reiterated what the Pentagon and NASA had told her: the Navy needed NASA's permission to test the women on government equipment, and NASA had told the Navy it had no interest in the project. Without the "requirement," as the Navy called it, they could not go forward. It had been difficult to get a more specific response, Cobb complained. Everyone blamed the decision on someone else and it had taken her two days to track down that meager answer. Couldn't a compromise be worked out? another congressman asked. "Can you get 'half your cake,' "—go ahead with the testing, even if it might not lead to space flight? Hart agreed that would be a good start.[36]

Already impressed with the women's case and attempting to be helpful, Congressman James Fulton, Republican of Pennsylvania, raised a pivotal point: "Did it strike the women that the reason the tests were canceled was because men thought the women were too successful?"[37] The audience and congressmen convulsed with laughter. It seemed ridiculous to suggest that women pilots might outperform the Project Mercury astronauts and just as ludicrous that the serious men at NASA would feel threatened by thirteen women, some of whom were housewives and mothers. As Fulton's suggestion dissolved into a few final chuckles, the subconscious impact it had on the subcommittee was obvious. Many members of Congress thought Fulton was a buffoon, and members of his own committee had heard too many of his bizarre ideas about space. Most members of the committee needed to hear only that Fulton supported an idea to decide against it. They did not take him seriously and readily dismissed his thoughts— even ones, like this one, that contained a measure of truth. Before Cobb and Hart could respond to Fulton's question, another congressman jumped in—almost to silence his ridiculed colleague. "Miss Cobb," he observed, "you showed a little bit of resentment toward" female monkeys who are training for spaceflight.[38]

As Cobb and Hart continued to answer the subcommittee's questions, Jackie Cochran made her appearance. She had been delayed by another meeting and had not heard Cobb's and Hart's opening statements. Anfuso halted the proceedings as soon as he saw Cochran enter the room. "Without question," he said by way of greeting, "the foremost woman pilot in the world . . . who holds more national and international speed, distance, and altitude records than any other living person."[39] After a few more exchanges, Anfuso called for a recess to conclude the morning's first round of testimony. While wary of what Cochran would say, Cobb and Hart were pleased that the subcommittee seemed at least to be favoring continued testing of the thirteen women.

Following a brief break, Congressman Anfuso called the meeting back to order. Cochran began her testimony with a statement that was both a strategic maneuver and a lie: "I only heard, Mr. Chairman, on Thursday, when I was out West, that I was going to be requested to come before your committee, and I had no opportunity to prepare very much of anything but my own thoughts more or less off the cuff."[40] Cochran used the old political trick of lowering expectations in order to impress the audience. She had, in fact, drafted her testimony nearly a month before and sent it out to many men at NASA, in the federal government, and in the military, and incorporated edits by Randy Lovelace and James Webb.[41]

Cochran told of her World War II experience—how she had organized the WASPs by selecting and training more than one thousand women to serve as pilots assisting the Army Air Force. She then moved directly to the question before the subcommittee, stating, "I do not believe there has been any intentional or actual discrimination against women in the astronaut program to date." Cochran used herself as an example: "As one who has had much experience in high speed precision flying and over the years has passed many of the tests that were given to select the seven first astronauts and also as one who would like exceedingly to go into space, I do not feel that I have been the subject of any discrimination."[42]

As she often did when she wanted to be forceful, Cochran numbered her points and shot them out like bullets. The subcommittee was asking the wrong question, she implied. The hearing should not be trying to determine if women had been discriminated against by NASA, but whether including women would "speed up, slow down, make more expensive, or complicate the schedule of exploratory space flights our country has undertaken." Let NASA determine how a women's astronaut program would affect the current launch schedule, she argued, adding, "There is no shortage of well-trained and long-experienced male pilots to serve as astronauts." Janey Hart squirmed in her chair. She'd known that Cochran would question why women should be included as astronauts if plenty of men were available. No sufficient scientific findings really existed that showed how women might compare to men in spaceflight, Cochran continued. A few women "might not be representative of women as a whole," she cautioned. "Based on my experience with women in the WASP program," she said, women "will prove to be as fit as men, physically and psychologically, for space flying. But such proof is presently lacking." Cochran then presented her recommendations, what she termed a "simpler and sounder way" to use women pilots. First, a large number of women—not necessarily pilots only—should be assembled for medical and scientific checks. "They should be well organized and supervised," she emphasized, sounding as though she were lobbying for the supervisor's job. With a properly structured program, Cochran argued, "a well-selected group of a dozen or more qualified women" might be ready to start an "astronaut trainee program for women." Jerrie Cobb and Janey Hart sat in the back of the room, bewildered and angry. Why was Cochran calling for a long research program that would yield a dozen women when thirteen women pilots already sat, ready and eager to begin training as quickly as tomorrow? Before taking questions, Cochran interjected one additional and potent comment: "Also, I have in mind the need for a large group considering the time the research will take and the natural rate of attrition among the volunteers due to marriage, childbirth, and other causes."[43]

Immediately questions focused on the high cost of training women who might soon return to the home. "We do not want to slow down our program . . . and waste a great deal of money when you take a large group of women in, because you lose them through marriage," Cochran said.*[44] Marriage, not sexism, was the reason women are not airline pilots, she declared. It takes over $50,000 to train a pilot to fly for the airlines, she stated. "That is expensive if you lose them through marriage."[45] A congressman agreed. He had been told by businessmen that this fact made it far more expensive to train women than men. That was why women were often paid less, he reasoned.[46] Motherhood and the space program did not seem to work together, Cochran inferred, yet women who were not mothers or not married also raised Cochran's suspicion. Adding a veiled invective toward single women and lesbians, Cochran stated, "I think first and foremost no one is successful unless they are first a woman and first a man and have all the instincts and desires of the two sexes."[47]

The subcommittee then asked for Cochran's opinions on jet test pilot experience. The test pilot requirement for becoming an astronaut is important, she said, explaining that it taught her a great deal about how planes are rigged.[48] But Cochran was well aware that only one woman could meet the test pilot requirement. "What woman except me could claim that on her résumé?" she often had joked.[49] Congressman Fulton pointed out that the current corps of male astronauts had gained their test pilot experience in the military. What if we opened the service academies to women? Then women could begin to receive training that eventually could lead to test pilot work? He asked Cochran directly: Would you open the Air Force Academy to women? Cochran instinctively sensed the subcommittee's silent derision toward the congressman. She digressed into a long recounting of her World War II experience. When Fulton asked the question again, Cochran bristled. "May I finish, sir?" she responded. "I can't get to your question unless you hear my thinking."[50] Cochran again recounted how she had organized the WASPs before eventually returning to Fulton's question. "We are in a new era," she stated. "I don't think you should open the [Air Force] Academy to

---

*Cochran's inference that male astronauts would make a career of their work at NASA was not borne out. In fact, John Glenn stayed with the astronaut program only two more years, and Scott Carpenter remained with NASA only until 1967. The same was true for many military pilots who left the service after their initial stints in order to fly for commercial airlines.

the women. Maybe never. . . . Don't clutter up the Air Academy with women unless we know we want them."[51]

The first day of hearings came to a halt at noon as Congressman Fulton launched into a convoluted and hyperbolic speech about the superiority of women and the heroic acts women had displayed through the ages. The maid of Orléans had led ten thousand men when no one else could, he argued. As subcommittee members rustled their papers in an effort to dash out of the hearing room for lunch, Fulton pleaded for their attention. "I am very serious about it," he insisted. Anfuso intervened, embarrassed by his colleague's rhetorical grandstanding. "Mr. Fulton is a bachelor," Anfuso told the audience, attempting humor, "and he thinks women are out of this world. He would like to get them out of this world."[52]

On Wednesday morning the hearings resumed. This time the mood in the committee room was heightened with excitement and anticipation. Flashbulbs popped as astronauts John Glenn and Scott Carpenter and NASA's George Low entered the room. The astronauts had just returned from the country's first manned orbits around the Earth and were greeted by the subcommittee as heroes. "Today we have with us two Americans of heroic stature, of whom nothing further need be said," Anfuso stated proudly. Congressman George Miller, chairman of the full Committee on Science and Astronautics, who did not attend the first day of hearings, cleared his schedule to be present for Glenn and Carpenter. Neither astronaut knew exactly why he was there. They did not know Jerrie Cobb personally, and had heard little about the other women's testing in Albuquerque and nothing about isolation tests or Pensacola. NASA did not coach the men on what to say and the astronauts had not prepared an opening statement. The subcommittee staff had instructed Glenn and Carpenter to answer questions based on their own experience. Later Glenn thought he understood the women's motivation in calling for the hearings. They "got the bit in their teeth and thought it should go forward," he said. But at the moment, the sole reason the men could think of for finding themselves seated at the witness table that morning was that they were the country's only two orbiters, a fact that put them in a unique position to discuss what qualifications an astronaut needed for future orbital flights. Glenn thought their instant celebrity had something to do with it as well.[53] With his new hero status, Glenn recently had been called upon to express his views on nearly every subject, even addressing a joint session of Congress about patriotism and the future of the country. George Low understood why he was there. Questions about women astronauts often ended up on his desk. Just last month, a staff member at NASA asked Low how he would respond to letters from women who wanted to become astronauts. "I hope you will have ideas on answering these letters," the NASA staffer wrote. "As you know, women fall within the physically qualified. There must be other valid reasons" why they are not in the program.[54]

Low opened the second day of hearings with a review of NASA's astronaut qualifications. The requirements had worked well for the initial Mercury program, he said, and were being reconsidered for the upcoming Gemini and Apollo programs where NASA objectives would focus on two-man missions and the lunar landing. No woman had ever been intentionally disqualified from consideration. In fact, NASA had about six women apply for the next round of astronaut selection. None of them was accepted, he admitted. How many women are test pilots in the United States? Anfuso asked. "Miss Cochran, Mr. Chairman, is the only woman test pilot that I know of." Anfuso did not seem satisfied and wanted reassurance from Low that the space agency was not discriminating against females. "We do not want to leave out the women." Anfuso warned. "No, sir; I certainly don't," Low agreed. Besides, it really was lack of interest rather than discrimination that caused so many women to shy away from scientific careers, Low argued. "I don't believe, Mr. Chairman, that there is any discrimination against women in aerospace engineering," Low offered as an example. That response relieved Anfuso, who readily accepted Low's rationale. "I'm glad you answered the way you did," the congressman said, "because this question was given to me by a woman who thought there was that discrimination." Congresswoman Weis, who sat with other members of the subcommittee listening to the exchange between Low and Anfuso, was not convinced. While there seemed to be no intentional discrimination against women in astronaut qualifications, she said, there was definitely a roadblock—a discrimination that might not be intentional—but was nevertheless built in. Low disagreed. "I see no reason why women should not enter into the test piloting field. I don't think that in the civilian test pilot area there are any roadblocks now. It is just that none of them have seen fit to get into this area," he said.[55]

There was another problem as well, Low continued. Women would be interfering with the current program if they wanted to use the equipment such as the centrifuges and vacuum chambers. "All of this equipment is very much loaded up at the present time," he said. "That is the best point you have made," Anfuso said enthusiastically. He tried to put Low's argument into his own words for the benefit of the subcommittee. "In other words, you are not objecting to women, but at the present time, to let them use the things that you are using now for the astronauts, would be interfering." Yes, that's it, Low said.[56] Even Jackie Cochran, who was not attending the second day of hearings since she was not required to testify, knew that the Air Force always seemed to find time to help her with her record-setting flights at Edwards. Three months earlier, Navy brass had told her that they had the facilities available for women; the Navy just needed NASA's green light.[57] Cathryn Walters, who sat next to Jerrie Cobb behind the witness table, disagreed with Low's assertion. As Dr. Shurley's assistant on the isolation tank experiments, she had been invited to observe the Navy's psychiatric facilities in Pensacola. She knew the doctors on base were interested in moving forward with the research and hoped to set up an exchange of scientific information with Shurley. In fact, Pensacola doctors seemed eager to be involved in the women's testing; it brought intellectual energy to their work which often was weighed heavily toward clinical practice. They had celebrated when Cobb passed her exams on the base.[58] Cobb also found Low's comment absurd. NASA seems to have enough time and equipment to train chimpanzees, she told herself.[59] Besides, the notion that women could only use equipment after the men were done sounded like the excuses she had heard back at Classen High more than ten years before. How many times had she heard that girls' athletic teams could not practice in the gym until the boys were finished?

Congressman Fulton looked concerned about the direction of questioning. It seemed to miss the point, he thought, that rules could be bent, had been bent, and should be bent, in order to find the best people to become astronauts. He put his sights directly on John Glenn. "On the basis of the requirements that Mr. Low has stated, obviously Colonel Glenn would have been eliminated. You wouldn't have passed, because you don't have an engineering degree, do you?" Fulton asked. Glenn, sensitive about his lack of a college degree when he was undergoing astronaut testing, quickly responded. "I have one now," he replied, relieved that Muskingum College had recently accepted his extension school and correspondence credits and awarded him a bachelor's degree.[60] Fulton was right: NASA had waived the college degree requirement for Glenn and Carpenter when they were selected as astronauts, accepting what it called their "equivalent experience." "We can't look at these methods of selection and requirements as rigid," Fulton argued. He did not find fault with either John Glenn or Scott Carpenter; he considered them outstanding astronauts—"you are tops," he later told them.[61] He believed that these men, who did not precisely qualify under NASA's set rules, proved his point: it would be necessary to make exceptions in order to find the best people.

Fulton turned to Scott Carpenter and pressed the point. Wouldn't putting the first woman in space be a worthy national goal? he asked. During these early stages, Carpenter countered, spaceflight was filled with un-knowns, and it was important to eliminate much of the uncertainty and danger before others took part. Fulton remained unconvinced. "It is the same old thing cropping up," he declared, "where men want to protect women and keep them out of the field so that it is kept for men." Agitated with what sounded like the start of another dramatic monologue, Anfuso tried to step in just as Fulton began swelling about Molly Pitcher, Queen Elizabeth I, Sacajawea, and Malinche leading Cortés. The audience began to laugh once more at Fulton's theatrics, and Anfuso worried that he was los-ing control of the hearings.[62]

John Glenn tried to redirect the testimony. "If we can find any women that demonstrate that they have better qualifications for going into a pro-gram than we have going into that program, we would welcome them with open arms." At that point the hearings completely erupted into laughter. Glenn's inadvertent gaffe brought down the house.[63] Scott Carpenter laughed as well, amused by the image of stalwart male astronauts opening their arms to receive women sent from Earth.[64] Glenn blushed and quickly tried to recover. "For the purposes of my going home this afternoon, I think that should be stricken from the record," he ventured.[65]

Glenn returned to the question of why some professions were dominated by men and others dominated by women. Glenn knew people who had chal-lenged those assumptions before. When he was a boy, he later said, his mother became the first woman elder in the Presbyterian Church for their region in Ohio. When an African-American soldier had wanted to go through flight training with the U.S. Navy, Glenn had witnessed the endless discussions about what integration would mean to white pilots. Some military men thought it would lower standards and dilute expertise. Glenn had substituted for the black pilot's flight instructor one day, and he had seen no reason why the man could not go forward.[66] People seemed threatened by change, Glenn believed. They wanted to hang on to what was familiar.

Glenn was philosophical—clearly more comfortable commenting on prejudice than actively fighting against it. What he said next defined the course of the hearings and would be remembered long after the subcommittee reached its final decision. "I think this gets back to the way our social order is organized, really," Glenn told the members of Congress. "It is just a fact. The men go off and fight the wars and fly the airplanes and come back and help design and build and test them. The fact that women are not in this field is a fact of our social order. It may be undesirable."[67]

Janey Hart could barely stay seated. "*Fact* of our social order," "*may* be undesirable"—her mind raced with heated rebuttals and challenges that she wanted to voice. People should not be content with the way things are. They should try to improve them. What if everyone had been content with slavery? she thought.[68] Inequity had to be challenged. There is no greater American hero right now than John Glenn, Hart said to herself. Why won't he use his position to show some leadership? Glenn's reluctance to attack the social order enraged her. She heard his comment as compliance and his unwillingness to fight discrimination as a failure of intellectual ability and guts.[69]

"I think Colonel Glenn has hit . . . on the exact differences of opinion that exist here," a congressman confidently interjected. Our social order dictates these differences and the space program merely follows the gender distinctions that time and history have prescribed. "I do not think the women of America want to do all the things that the Russian women have to do." He declared that he saw no designs to keep women out of the space program. "That has not been an intentional thing by any stretch of the imagination."[70]

Fulton asked one final question of John Glenn: Would you support a program to train women astronauts? "I wouldn't oppose it," Glenn replied, but then he added, "I see no requirement for it." Glenn argued that NASA already had many qualified men and had spent a lot of money training them. "Now, to spend many millions of dollars to additionally qualify other people, whom we don't particularly need, regardless of sex, creed, or color, doesn't seem right, when we already have these qualified people."[71]

The second day of hearings had been a nightmare for Janey Hart and Jerrie Cobb. They were relieved to have one more day to address NASA's damaging testimony. Anfuso, seeing that it was almost noon, brought the day's hearings to a close. He turned to John Glenn and Scott Carpenter and congratulated them on their great achievement and leadership, noting all the mothers in the audience who had brought their children to the hearings just to see the astronauts. Their leadership was the kind everyone wanted and respected, he said. He banged the gavel and announced that the hearings on astronaut qualifications were concluded. "This may be the last time that I will preside over a committee," he said, referring to his impending retirement from the House. Acknowledging the accolades of fellow members of Congress who saluted him as he set out on a new career, Anfuso thanked the committee for its "perfect harmony."[72]

Jerrie Cobb was stunned. Everyone around her was getting up and heading for the doors, seemingly oblivious to what had just happened. Was it over? Wouldn't there be a third day of testimony as Anfuso had promised? What about the time for rebuttal that she had requested in her opening statement? Cobb turned to Jane Rieker and Cathryn Walters, who were sitting next to her, but they could only shake their heads. They knew that the hearing had turned against Cobb from the moment Jackie Cochran began her testimony. Glenn, Carpenter, and Low had underscored what the subcommittee seemed to have already decided. But Jerrie Cobb was unwilling or unable to give up. As she always did, even in the face of overwhelming argument against her, Cobb reasoned there must be another chance. That evening after Cobb, Rieker, and Walters returned to their Washington hotel, Cobb received official word that Anfuso believed he had enough evidence for his report, and another day of testimony was not necessary. Anfuso did offer Cobb a small measure of hope when he agreed to accept for-the-record written comments from anyone who had testified. Cobb and Cochran were the only witnesses who obliged. Cobb feverishly compiled a statement that called into question the absence of expert witnesses at the hearing. Then she gathered all her medical data—a handwritten chart from Randy Lovelace comparing her scores to the men's, Dr. Shurley's summary of her isolation tank experiment, the list of Pensacola tests she had taken. "All we need is the opportunity to prove we are 'capable,' 'qualified,' and 'required,' " she wrote, echoing Jackie Cochran's words. Meanwhile, Cochran's statement emphasized that there was not, nor had there ever been, an official NASA-sanctioned women-in-space program. She then challenged Cobb's position as spokesperson for the group and included Gene Nora Stumbough's letter as proof that some of the Mercury 13 disagreed with opinions Jerrie Cobb expressed.[73] The day after the hearings, newspaper accounts made the outcome clear. "Gently but firmly, a couple of American space heroes today drained the fuel from the proposal to train women astronauts," the *Chicago Tribune* reported. Although they had little reason to be optimistic about becoming astronauts, the rest of the Mercury 13 clung to one final hope—that the subcommittee's formal recommendations to the House Committee on Science and Astronautics might at least allow them to continue testing.[74]

THAT WEEKEND ATTORNEY GENERAL ROBERT KENNEDY, NEEDING A break from the Washington summer heat, traveled to Cape Cod and the Kennedy family compound on Hyannis. Before he departed, he called John Glenn and asked if he and his family would like to join him for some water skiing. Glenn readily accepted and was thrilled to discover upon arrival at the Cape that the President and First Lady also were there. Jackie Kennedy and John Glenn became the stars of the weekend's waterskiing show, crisscrossing in the bay and balancing on one ski as Ethel Kennedy steered the racing boat past the President's yacht. The weekend was magical, onlookers said. The President thought Glenn showed the makings of a politician.[75] That same weekend, Jerrie Cobb headed to Vermont, where she spoke to a Girl Scout Roundup. Girls swarmed around Cobb, taking her picture and peppering her with questions about the Lovelace testing. Some of the girls, inspired by Cobb's talk, built their own mechanical rocket and tested it for launch. "Space Dancers," they renamed their group. Among the two hundred girls who listened to Cobb, there did not seem to be any lack of interest in spaceflight.[76]

When it came to access to the President, Cobb could not compete with John Glenn. Over the next three months, her repeated requests for an appointment to see Kennedy were bumped to James Webb. They each arrived with a routine interoffice routing slip indicating the White House's official response: "An appointment with the President is not possible."[77]

In October, the House Committee on Science and Astronautics issued its annual report, including the recommendation from the subcommittee on astronaut qualifications: "After hearing witnesses, both Government and non-Government, including Astronauts Glenn and Carpenter, the subcommittee concluded that NASA's program of selection was basically sound and properly directed, that the highest possible standards should continue to be maintained, and that some time in the future consideration should be given to inaugurating a program of research to determine the advantages to be gained by utilizing women as astronauts."[78] The report's single sentence hit Jerrie Cobb like gravity.

CHAPTER ELEVEN

Afterburn

ALTHOUGH CONGRESS HAD ISSUED ITS DECISION, ARGUMENTS STILL
flared as Jerrie Cobb, Jackie Cochran, and NASA launched accusations and
mocking rebuttals. Cobb told the Mercury 13 to "break loose now" and take
a public stand urging the country to send an American woman into space.[1]
Either our "group is going to work together and support the program or
Cochran will make it a debacle," she warned. "I really feel sorry for Cochran
as she must be a pretty unhappy person." Later Cobb would add the words
"domineering" and "dangerous" to her assessment.[2] On the other side, a
NASA psychologist pronounced a women-in-space program to be of little
consequence and ridiculed Cobb's scientific arguments. Sure, women might
weigh less, he argued sarcastically—if you could persuade them to leave
their purses behind. "As for the ladies' alleged ability to withstand boredom
and confinement better than the man," he said, "I think there might be a
number of harried husbands who have sat through long evenings listening
to the wife's recitations of the day's activities, who have credentials in the
area of tolerance and boredom."[3] Other members of the Mercury 13 scruti-
nized the congressional transcript and joined the debate. Schoolteacher Jean
Hixson sought out a fellow WASP for her interpretation of Jackie Cochran's
remarks. "Cochran's arguments were about what I expected and did more
harm than good," her friend wrote. "Certainly all that dribble about getting
married and having babies didn't help the record!"[4] For her part, Jackie
Cochran took to denouncing Cobb's claims that she was the country's first

woman astronaut. Citing Cobb's entry in *Current Biography*, Cochran asked NASA: Isn't an "astronaut" someone who has actually flown in space or who is selected by NASA to fly?[5] Even Cochran's husband got into the fray, compiling a stinging list of "interesting questions for someone to ask Cobb."[6] Of all the charges that were fired back and forth, none was more repugnant than the one repeated by Wernher von Braun and attributed to NASA's Robert Gilruth, director of the Manned Spacecraft Center. Von Braun found amusing Gilruth's idea that women in space could be used sexually by male astronauts. "Another question that I am frequently asked is this," von Braun told a college audience: "Do you ever plan to use women astronauts in your space program?' Well, all I can say is that the male astronauts are all for it. And as my friend Bob Gilruth says, we're reserving 110 pounds of payload for recreational equipment."[7]

Amid the verbal scrimmages, the isolation tank testing that had already been lined up for the rest of the Mercury 13 never materialized. Dr. Jay Shurley's groundbreaking research on Cobb, Rhea Hurrle, and Wally Funk was stored away in his files and remained there for the next forty years.[8] The Pensacola physicians were equally disappointed. Like the other women, Jean Hixson returned the airline ticket to Florida she had bought, conceding that more testing was not likely to take place.[9] Hixson never heard from Dr. Lovelace again. One of Randy Lovelace's final written comments about the women astronaut project was made in a letter congratulating Jackie Cochran on her congressional testimony. "There have been many favorable comments in Washington about your testimony in contrast to other testimony," he wrote. The hearings either had convinced Lovelace that NASA would not sanction his project or Cochran had persuaded him that long-range rather than immediate testing was a better way to proceed.*[10] Apparently Jerrie Cobb and Janey Hart heard only silence from the Lovelace Foundation.

Only one scientific report was ever published on the Lovelace tests. It appeared in 1964 in the *American Journal of Obstetrics and Gynecology*. Two Lovelace Foundation physicians theorized that women's menstrual cycles

---

*Randy Lovelace told Jackie Cochran that the Lovelace Foundation had not determined any overall rankings for the thirteen women. Prompted by Cochran, Lovelace also told her that to his knowledge Jerrie Cobb had never been asked to be a spokesperson for the group.

might complicate their ability to work in space since "objective studies" had shown menstruating women to be inattentive and more prone to accidents. Citing previous research, they noted that "mental illness is higher, crime rate increases, and there are more attempted and successful suicides just prior to and during the menstrual flow."*[11]

Cobb continued pressing James Webb at NASA, hoping that he would change his mind despite the House subcommittee decision to discontinue testing. Webb would not budge and told Cobb he could not give her the kind of commitment she wanted.[12] Increasingly, Cobb's letters sounded more like pleading than persuasion. "You know that *nothing* is more important to me," she wrote Webb. "I beg of you, just for the opportunity to prove myself."[13] Cobb also started speaking publicly, without her self-imposed muzzle, since she believed she had little left to lose. Assuming she was still working as NASA's consultant on women's issues because she claimed the space agency had never officially informed her she was not, Cobb called herself the "most unconsulted consultant" in the U.S. government. In speech after speech, she used the phrase and sarcastically claimed that male scientists thought the ideal woman astronaut would be a rare breed—a smart, petite female from high altitude—or, as she put it, a "midget woman from the Andes Mountains with a Ph.D."[14] By the end of the year, Webb lost his patience and agreed to meet with Cobb in Washington to set matters straight. Since he felt Cobb often misinterpreted his comments or saw them as more encouraging than he intended, Webb asked a NASA assistant to take notes on what transpired. He told her that her NASA consultancy had expired and had not been renewed. He asked Cobb to stop using the "unconsulted consultant" phrase to refer to her current work since it gave the impression that she was still affiliated with NASA. In what almost sounded like a threat, Webb informed her that if he continued to be asked about her critical remarks or what he called the "irrational manner" of her campaign, he would be forced to reply that he did not consult Jerrie Cobb because he did not find her judgment useful.[15] As far as NASA was concerned, the case of women in space was closed.

---

*The authors also reported that the Civil Aeronautics Administration once warned that it was dangerous for women to fly planes three days before, during, or after their menstrual periods.

Cobb certainly felt that way when James Webb appointed Jackie Cochran as NASA's new special consultant.*[16]

For many of the Mercury 13, it seemed as if nothing tangible resulted from their testing, the subsequent political advocacy, or the congressional hearings. Some women resigned themselves to having been mere laboratory rats; others added new aviation ratings to their résumés in an attempt to prove NASA and the United States wrong.[17] Jerri Sloan resolved to speak out every time she witnessed NASA favoring white men or making snide remarks about women. She wrote letters to NASA, spoke on radio programs about her experience, conducted interviews with the American, Mexican, and British press. She would never forget the "110 pounds of recreational equipment" comment—it represented to her everything reprehensible about NASA's intractable sexism. Sloan went back to work at her air services firm, married her business partner several years later, and became a self-professed "guard dog," not as dynamic a role as being a WASP—Sloan's ideal of an active, patriotic woman—but one that might eventually cause NASA to choose a woman astronaut.†[18]

Wally Funk took a different tack. Unwilling to give up the dream of becoming an astronaut, she searched for unofficial ways to continue testing, hoping it would help in the unlikely event that she ever got the call from NASA. Funk found a few Marines at El Toro Marine Corps Base in California who allowed her to undergo a high-altitude-chamber test and the Martin-Baker seat ejection test. At the University of Southern California, she convinced scientists to put her in the centrifuge to measure her ability to withstand increasing gravitational forces. Funk knew that military personnel who took centrifuge tests wore G suits that tightened their muscles and helped them avoid blacking out. As a civilian, Funk was not permitted a government-issue

*Cochran's proposals for a women-in-space program also were met with disinterest by NASA. She sensed that her appointment might be more window dressing than substantive and several years later complained to NASA that it had done little with her suggestions and had rarely consulted her.

†One somewhat strange result of the women's testing was a job offer that Jerri Sloan received. After Sloan's photograph appeared in the press as one of the women who had taken astronaut tests, the DuPont company thought she had modeling potential. DuPont asked if Jerri would tour the country and appear in advertisements publicizing their new fabric, Lycra. "With kids to feed," Sloan Truhill later said, she accepted the offer and toured the United States and Mexico for several months.

G suit. "I asked my mother if she could give me her worst merry widow," she said, describing a girdle that cinched a woman's figure into an hourglass shape. Constricted by the "Wally G suit" under her bulky flight gear, she climbed into the centrifuge, began whirling around, and soon felt the "gray curtain" of unconsciousness fall upon her. She clenched her muscles and hoped that the merry widow would keep "everything onwards and upwards." Funk reported that USC personnel were amazed to discover someone who could withstand five Gs without the aid of a G suit. The scientists would have been even more shocked had they realized she experienced the run wearing a corset. Funk had not understood precisely that G suits function by squeezing one's lower extremities, not the midsection. The merry widow could have done her significant harm, constricting Funk's ability to breathe deeply and restricting blood flow into her chest. In fact, her successful run on the centrifuge occurred in spite of Funk's misguided use of a corset, not because of it.[19]

While the immediate results of the Mercury 13's efforts were negligible, there were profound long-term consequences. The first came in a phone call to Janey Hart's home in Washington. It was Betty Friedan. She had read Hart's testimony before the congressional subcommittee and thought it was wonderful—incisive and principled, even brave. She had just published a book on the stifling lives of white suburban housewives in America called *The Feminine Mystique.* People everywhere were talking about it, Hart knew. She had read it, and the book had made a strong impression on her. Friedan argued what Hart had already sensed—that women were viewed as second-class citizens by the media, education, the government, industry, religion, and nearly every other facet of American society. Friedan asked if Hart could get away for a day to go to New York for a meeting. Maybe there's something we could do together, Friedan told her; maybe women need an organization, something that would pick up where the suffragists left off, maybe even a revitalized movement. Hart channeled her disappointment with NASA into that meeting and the many others that followed. In 1966, the National Organization for Women (NOW) was founded with Friedan serving as its first president and Janey Hart as a member of the organization's first national board. Among NOW's first actions were forcing newspapers to eliminate sex-segregated help-wanted ads and filing a formal complaint against NASA that charged discrimination in the hiring not only of astronauts but also for top-level posts in its administration.

The attention Cobb and Hart brought to NASA's astronaut-hiring practices did not precipitate immediate results but raised related issues.[20] Edward Dwight, Jr., an Air Force captain, was among the first African Americans to go through jet test pilot training at Edwards. Dwight did well and in 1963 was one of twenty-six pilots the Air Force recommended to NASA for astronaut service. NASA rejected Dwight, a decision many of Dwight's supporters believed had more to do with racism than with any weakness in his flying credentials. Throughout his time at Edwards, some officers viewed Dwight as a "Kennedy boy"—a pilot who had been admitted to test pilot school only because President Kennedy had called for racial integration. Although Dwight met every astronaut qualification, including jet test pilot experience and an aeronautical engineering degree, the Air Force subsequently assigned him to a desk in Ohio. Dwight's protests eventually stirred media attention and prompted congressional inquiries, and even a response from the Soviet Union. The TASS news agency reported that Dwight "was rejected for astronaut duty because he is a Negro." NASA claimed it had a perfectly good record on equal opportunity, but observers, who were now aware of the circumstances of Captain Dwight as well as those of the Mercury 13, were beginning to wonder.[21]

NASA had more questions to answer on June 16, 1963, when Soviet cosmonaut Valentina Tereshkova became the first woman in space. To Cobb, the news that a female textile worker and amateur parachutist had beat her into space was demoralizing. Tereshkova was not even a pilot, and she certainly did not have an engineering degree or jet test pilot experience. "I've fought the battle so long, I can't help feeling a little regret," Cobb said. "I know . . . we could have done it. Now we've lost our only chance to have a first in space. . . . I really mean it when I wish her well," Cobb admitted. "I'm glad a woman made it. But I'm sorry she's not an American."[22] John Glenn spoke out as well and maintained his belief that American men made better astronauts than American women would. The qualifications we're looking for are best met by men, he said.[23] Celebrated writer Clare Boothe Luce wrote a scathing article in *Life* magazine, reminding readers that a year earlier, thirteen American women had asked Congress to send an American woman into space. Where are those thirteen women now? Luce asked. "The U.S. Team Is Still Warming Up the Bench" the *Life* headline answered. Luce called the missed opportunity a costly Cold War blunder and excoriated American men for their sexist views of women. NASA disagreed, calling Tereshkova's flight nothing more than a publicity stunt.[24]

Hoping that James Webb at least might be chagrined to see another space record go to the Russians, Cobb formally applied to the NASA astronaut training program a few weeks later. NASA rejected her application outright, stating that it had come in after the deadline and would not be considered. A NASA spokesman added that two other women managed to get their applications in on time. Of course they, too, would later be rejected.*[25] In November, Cobb flew to Jamaica, a place she fondly remembered from her early days ferrying planes, to consider what she should do next. She knew she would have to leave her marketing position at Aero Design in Oklahoma City. As much as Cobb had benefited from the unlimited support of her boss, Tom Harris, she could not put on a smile and greet the public anymore.[26] While she wanted desperately to go into space, she also realized that her public campaign had come at a high personal price. She needed to return to the more private Jerrie and decide how she wanted to use her flying ability, her time, and her life.

In November 1963, the assassination of President John Kennedy convulsed the nation, and suddenly Lyndon Johnson found himself in the Oval Office deciding if he would follow Kennedy's space priorities or change them. Just days after the funeral, an exhausted Johnson called James Webb on the telephone. After briefly discussing a proposal that the military take over the Gemini program—trial runs for the Apollo moon program—Johnson asked how the future space budget looked. There was not as much money as he had hoped, Webb replied. He wondered if Johnson were prepared "to slip the lunar landing out of this decade."[27] But a man on the moon was one Kennedy pledge no one was willing to see lost. Kennedy's goal for space had defined his presidency's vision and vigor. Johnson wanted the lunar launch to go forward.

As NASA continued to push ahead with the Gemini schedule it also brought more specialists into the work.† In an April White House ceremony,

---

*Cobb also applied for the high-altitude X-15 test pilot program and was turned down. "Essentially, this conclusion is based on the same factors which have precluded you from service as an astronaut," NASA informed her.

†The goal of the Mercury program was to orbit a man around the earth. The Gemini space program sought to orbit two men over a period of several days, during which they would perform rendezvous and docking maneuvers. The Apollo program focused on the lunar landing.

Lyndon Johnson named Dr. Randy Lovelace the new director of space medicine for manned spaceflight at NASA. Lovelace threw himself into work, his pace more frenetic than usual. He would work in his Washington office several days a week, then return to Albuquerque to perform surgery and oversee operations at the foundation. After having fifty-seven formal speaking engagements in one year, in addition to his regular work schedule, Lovelace realized he needed to spend more time with his family. He went on a round-the-world trip with his wife and two youngest daughters and put down money on a new vacation condominium in Aspen, Colorado.*[29]

In December 1965, Randy Lovelace hired a corporate pilot for the short trip to the condo, as was his custom. It was a clear day for the return flight to Albuquerque, and the twin-engine Beechcraft rose softly above Maroon Creek so Lovelace and his wife could look down on their new vacation home. The plane circled back east over Independence Pass toward the crest of the Sawatch Range. Realizing too late that he had swung into a tight box canyon without room to turn around, the pilot made a desperate steep turn before the plane hit the sheer rock wall, tumbled, cartwheeled, and crashed into the snow below. It took days before eighty planes covering 30,000 square miles could locate the plane and the three bodies. When rescuers finally discovered the crash site, they found the pilot dead and leaning against the fuselage. Still gripped in his hand was the plane's radio—its cord disconnected and blowing in the mountain wind.[30]

The memorial service for Randy and Mary Lovelace attracted hundreds of people, including generals, politicians, eminent scientists, and NASA personnel. Scott Carpenter represented the astronauts whose careers had begun in the halls of the Lovelace Foundation. Jackie Cochran was devastated by the loss of two of her closest friends.† In eulogies that poured in from Washington and Cape Canaveral, many people made note of the poignant

---

*Randy Lovelace also was thinking about writing his autobiography and brought many of his space files home so that he could consult them. Lovelace's youngest daughter, Jacqueline Lovelace Johnson, believes those files have been lost.[28]

†Immediately following the deaths of Randy and Mary Lovelace, family and friends wondered if Jackie Cochran might assume parental responsibilities for her goddaughter, sixteen-year-old Jackie Lovelace, but the girl eventually went to live with an aunt and uncle. Cochran kept in close touch with Jackie Lovelace and her older sisters. Looking back, Jackie Lovelace Johnson said that Cochran was a tough woman in a tough time and did not have a lot of compassion. Compared to Jackie Cochran's ego-driven determination, everyone else was "a piker," Johnson said.

timing of Dr. Lovelace's death. At almost the exact moment that the bodies were discovered, astronaut crews from *Gemini VI* and *Gemini VII* orbiting high above the earth completed the first true space rendezvous, flying within a few feet of each other in practice for the lunar landing. From their position in outer space, the astronauts could almost see the jagged mountainside where Lovelace's plane had gone down. President Johnson, in commenting on the legacy of Randy Lovelace, said "his life was too short, although his legacy to space medicine will endure and will be a resource of assurance to future astronauts whose names and deeds are yet unknown."[31]

Jerrie Cobb had to wonder what might have happened to the women-in-space program had Randy Lovelace lived. Given his new position at NASA, Lovelace might have been able to reintroduce the testing of women at a later time and as an insider. Some of Lovelace's colleagues in the aviation medicine community, such as Dr. Stanley Mohler, thought that Randy Lovelace always seemed to find a way to maneuver around bureaucratic and political obstacles to eventually get what he wanted. His scientific curiosity alone made Lovelace continue to investigate questions that other scientists thought were not worth pursuing. To many, it seemed as though Randy Lovelace would probably have waited a bit before pushing for women astronauts again.[32]

After leaving her job at Aero Design in Oklahoma City, Jerrie Cobb settled in Florida, where she shared a home with Jane Rieker and tried to come to terms with what had happened to the women-in-space program. Cobb and Rieker began work on an autobiography chronicling Cobb's campaign for women astronauts and her early life as a record-breaking pilot. Both women hoped that *Woman into Space: The Jerrie Cobb Story* might revitalize the public's interest in Cobb's dream of becoming an astronaut, but there was no groundswell of support. Cobb then became an "aviation consultant," a term she considered embarrassingly grandiose. "I'm actually just a general flunky, working on my own. I'll fly anywhere for anybody who needs something or someone delivered," she said.[33] The closest Cobb came to the space program was flying photographs and magazine copy about a recent Gemini launch to the Chicago offices of *Life* magazine. Over time, she found herself flying more and more to Central and South America. She was familiar with the air routes from her days flying with Jack Ford and Fleetway, and she continued to be drawn personally to the green jungles, the deep forests, and the indigenous peoples of the Amazon. Perhaps more than any other attraction,

Cobb was taken with the solitude. She also immersed herself in religion. During one trip to South America she became interested in the work of the Wycliffe Bible translators—missionary translators who had lived with the native tribes, developed an alphabet for their unwritten languages, and prepared versions of the Bible. Quiet as she was about her religious faith, Cobb's friends knew that it was a source of great strength to her, especially as she sought to live with the disappointment of not becoming an astronaut. Cobb hoped to find a little freelance flying in South America that would allow her to devote the majority of her time to the indigenous tribes in the jungles. "I have a great love for Latin America and the people there," she said.[34]

After being turned down for several jobs as a pilot for organized Latin American ministries (Cobb believed she had been rejected because the groups were looking for male pilots), she decided to become her own one-woman missionary flight operation.[35] Accepting modest financial contributions where she could find them, Cobb spent her days flying alone above vast areas of the Amazon, taking food, medicine, and seeds to the people of the rainforest. At night she would consult her handmade maps and search for a landing strip, often a small, rutted clearing hacked out of the jungle with machetes. Cobb knew her family wondered if she was running away from her disappointments.[36] She also realized her friends worried about how long she could live in such an unforgiving environment, where it could take half a day to fill her plane's tank with gas—backbreaking work hauling five-gallon buckets and straining the gas through a worn shirt. But she was content, Cobb said. She had found a way to earn just enough money—sometimes performing such menial tasks as flying a case of Coca-Cola and a couple of bags of cement from one place to another—to keep her plane fueled and herself above the clouds.[37] Her battles with Jackie Cochran, James Webb, and politicians on Capitol Hill were far behind her. Cochran was having health problems and had to restrict her flying and professional commitments. Even Webb had moved on, resigning his post in 1968 a few weeks before Richard Nixon's election to the White House.* Cobb no longer wanted to face man-made chal-

---

*Webb believed he was viewed as a partisan Democrat and that a more politically neutral leader at NASA could keep the lunar landing on schedule. Webb also became a target of NASA critics after the 1967 Apollo fire that killed astronauts Gus Grissom, Roger Chaffee, and Edward White. Webb later became a regent of the Smithsonian Institution. He died in 1992.

lenges. She much preferred confronting natural ones such as rain and fog and even a jungle pilot's worst problem—running out of daylight. With no twilight in the Amazon, day would turn to night in an instant. Now literally living in a world of day and night, black and white, life and death, Cobb was restored by an environment seemingly devoid of ambiguity. "When I first arrived in Amazonas," she wrote in her autobiography, "I was not sure I was brave enough to fly over such an immense jungle." Now she did not know if she had the courage to leave.[38]

Flying high above the Amazon on July 20, 1969, Cobb heard the news over the plane's radio that Neil Armstrong had set foot on the moon. "That's one small step for man, one giant leap for mankind," Armstrong said as he eased down from the module onto the powdery lunar surface. Kennedy's mission had been accomplished. Cobb was thrilled to hear the report; anything to do with spaceflight still excited her. But she also realized that the country's single-minded commitment to beating the Soviets to the moon had pushed every other space accomplishment aside. In many ways, John Glenn had been right. It *was* a fact of our social order: women simply were not central in the nation's priorities and a few individual women were not going to convince the United States government otherwise. Hearing the historic news of the moon landing from the crackly voice of a radio operator in Peru, Cobb realized how dramatically her life had changed. Eight years earlier, when Alan Shepard had made the first launch, Cobb had been at the very center of the excitement—responding to reporters' questions, appealing to James Webb, petitioning the Vice President, testifying before Congress, attending an endless round of dinners and speeches with pearls, high heels, and smiles. Now she was alone, far above the jungle, with nearly no one to share in her excitement or her regret.[39]

With the return of the *Apollo 17* mission in December 1972, NASA completed the eleven-year, $25.4 billion program whose singular goal had been human exploration of the moon. In all, twelve astronauts had walked on the moon during six lunar-landing missions. Starting in 1970, even before the completion of Project Apollo, NASA began to experience severe cuts to its budget. Given the political climate at the height of the Vietnam War, Congress sensed that the American public was far more concerned about domestic and international problems than the exploration of space. After Apollo, NASA focused its efforts on extended human voyages in space. Its first experiment with long-duration spaceflight was the 1973 Skylab program, which launched three manned missions. In the waning days of the space race with the Soviet Union, NASA's goals shifted to increased cooperation in space. In 1975, NASA and the Soviet Union joined together for the Apollo-Soyuz Project. Later, the U.S. space agency developed scientific spacecraft for planetary exploration, including the Viking project, which searched for signs of life on Mars.[40]

Social change propelled by the women's movement and the civil rights movement also had an effect on NASA. Amendments to the 1964 Civil Rights Act put federal hiring policies under scrutiny. One by one the armed services opened military flight training, including jet test pilot schools, to women.*[41] The NAACP and the Urban League pressed NASA to explain why it selected only white male astronauts. The United States Civil Rights Commission asked NASA for gender and racial data regarding its astronaut-selection process. NOW picketed outside NASA headquarters in Washington and staged street theater protests to garner press attention.† The Senate Space Science and Veterans Committee held hearings to investigate NASA's Equal Opportunity Office.[42] Even George Low, who had defended NASA's hiring policy at the congressional hearings eleven years earlier, was now willing to admit that its record for bringing women into upper-level positions was "very poor."[43] Congresswoman Barbara Jordan went further. She argued that the subtle and open discrimination aimed at women and minorities had deprived NASA "of a vital resource for talent and ideas."††[44]

---

*Women were not allowed into the ranks of military jet test pilots until after the feminist activism of the late 1960s and early 1970s forced American culture to reevaluate its gender assumptions. The Army opened its test pilot training to women in 1974. In 1983 the Navy followed suit. The Air Force did not open its doors to women test pilots until 1988.

†One of NOW's protests followed Neil Armstrong's walk on the moon and featured five women carrying large cardboard cutouts of feet, each toe painted with red polish. When the five protestors stood in line, the words on their cutouts invoked and amended Armstrong's famous words. The NOW protestors' five feet read "One Giant Leap for Humankind."

††NASA also sought to improve its image among women and minorities and hired African-American actress Nichelle Nichols to publicly promote careers in space and write a report on NASA's recruitment policies. Nichols, who produced motivational films for minority youth, was best known for her role as Lieutenant Uhura, communications officer for the starship *Enterprise* on the popular television series *Star Trek*. Nichols's assessment of NASA's approach to hiring minorities and women was especially astute. By taking the "path of least resistance," she said, NASA had decided it was simply easier to put a man on the moon than to address issues of human equality.

Amid growing public and government pressure, the new administrator at NASA, James Fletcher, called for a different approach for selecting the next round of astronauts. He announced that "full consideration" would be given to minority groups and women for the large group of astronauts needed to fly upcoming space shuttle flights. In 1978, NASA selected thirty-five new astronauts from a pool of 8,079 applicants and introduced them to the public at the Johnson Space Center.[45] They included NASA's first African-American, Asian-American, and women astronauts. The six women, all mission specialists, were selected for their scientific expertise and were not pilots or commanders who would actually fly the shuttle. Selected as NASA's first women astronauts were Anna Fisher, Shannon Lucid, Judith Resnik, Sally Ride, Margaret Seddon, and Kathryn Sullivan. But Sally Ride was the one who would make space history. When she lifted off the launch pad on June 18, 1983, and became the first American woman in space, a crowd of half a million women and men lined the highways and beaches of Cape Canaveral to cheer her on and celebrate the crossing of a new national milestone. Long-forgotten by the crowds awaiting the countdown were the Mercury 13, Randy Lovelace, and even Jackie Cochran, who had died of heart disease three years earlier, at age seventy-four. Only Janey Hart was on hand to watch the shuttle lift off. Hart could hardly believe the crowd's exuberant roar as the shuttle slowly climbed into the sky. "Ride, Sally Ride!" they yelled.[46] While it appeared to nearly everyone else looking up into the summer sky that the nation's first woman astronaut was being hurtled into outer space on two massive booster rockets, Janey Hart saw it differently. She knew Sally Ride had been launched two decades earlier by thirteen American women and the dream of spaceflight.

Left Seat

EXCITED AS SHE WAS ABOUT SALLY RIDE'S LAUNCH, JERRI SLOAN Truhill knew that one more NASA threshold remained to be crossed: the left seat. Truhill wanted to see a female commander of the space shuttle, a woman leading a mission into outer space. Even after 125 missions spanning four decades, NASA still had not selected a woman commander. Sally Ride had been a mission specialist, a scientist who carried out experiments and research aboard the shuttle. She did not have any responsibility for flying the spacecraft "or any of the fun, either," Sarah Gorelick Ratley added. "We want to see a woman driving the bus, not sitting in the back," Truhill said. Truhill had been in touch with her Lovelace testing partner, B Steadman, as well as Janey Hart and some of the other Mercury 13 women. They vowed to keep speaking out publicly until NASA tapped a female commander. They saw a woman in the left seat as the personal fulfillment of their dreams and their lasting legacy.[1]

In 1998, President Bill Clinton announced that NASA had selected Eileen Collins to become the first woman to command the shuttle. A forty-two-year-old Air Force lieutenant colonel, Collins was a seven-year veteran of the astronaut corps and the second woman to go through military jet test pilot training at Edwards Air Force Base. She became a test pilot in 1990, one year after the Air Force permitted women to train at Edwards. In a White House ceremony celebrating her appointment, Clinton pointed out the contrast between Collins's selection and the public presentation of the Project

Mercury astronauts. "Forty years ago," he said, "*Life* magazine introduced America's first astronauts to the world, noting that the seven Mercury astronauts were picked from 'the same general mold.' They were all military pilots. They were all in their thirties. They all had crew cuts." With the appointment of Eileen Collins, the President said, the United States had crossed a milestone in space history and social equity.[2]

Collins knew she had not achieved the feat alone. She had been an avid reader as a child and had been inspired by books about Amelia Earhart, early women barnstormers, and the WASPs during World War II. For years she had been a "car hood" pilot—sitting on the family automobile in Elmira, New York, and watching planes take off and land at the local airfield. With limited funds and often relying on food stamps to see them through, the Collins family knew flying lessons were an extravagance they could not afford. Collins began working as a waitress at a local pizza joint, saving her tips for classes at the community college. When her bank account reached $1,000, she brought all the cash to a nearby flying school and earned her pilot's license. Then came Syracuse University, service in the Air Force, and jet test pilot training at Edwards. The reading, family support, public access to education, and open doors of the military all made a difference, Collins said.

Eileen Collins had been barely five years old when Jerrie Cobb and the Mercury 13 passed the Lovelace tests and pressed Congress for a chance to be launched into space. Collins never read the articles about Cobb floating in the isolation tank, or learned about Janey Hart's indictment of NASA or Wally Funk's merry widow ride in the centrifuge. When she joined NASA, Collins heard vague stories from the other women astronauts about a gutsy group of women pilots who had gone through testing a generation earlier. The physical and mental tests these women had to endure were more arduous than what current astronauts face, her colleagues told her. No one could recall many more details about them, and Collins did not know who the women were or what they actually did for a living or even if they were still alive. Then she received an invitation to meet them. More than thirty years after their initial testing, the Mercury 13 women gathered in Oklahoma City for a celebration of their own achievement and a salute to Eileen Collins. Gene Nora Stumbough Jessen had organized the gathering. An American woman was finally going to be flying a spacecraft, Jessen thought, and it was time for the space pioneers to see what their struggles had helped achieve.

It was not the first time some of the Mercury 13 had gathered. In 1986, B Steadman, cofounder of the International Women's Air and Space Museum, noted that the year marked the twenty-fifth anniversary of the Lovelace testing. She tracked down all the Mercury 13 through their membership in the Ninety-Nines, and a handful of the women traveled to Ohio to meet one another for the first time, while others—perhaps still disillusioned with their experience—remained at home. Irene Leverton was not sure she wanted to go. Over the years she had encountered so many "crummy deals," as she called them, in the aviation industry that she was did not know if she wanted to relive another one. She remembered being paid less than male pilots who performed the same work she did, being blamed for errors that inexperienced male pilots had committed, finding out she had been hired as a pilot because her employer could count her twice for official documents— once as a captain and again as a stewardess. Once, after a sadistic flight examiner tried to frighten her by rolling the plane upside down until Leverton dangled in midair from a loose seat belt, she grabbed a wrench from a hangar worktable and took off after him. Why would she want to think again about the disappointing astronaut tests? But she decided to go to B Steadman's reunion anyway. When she was asked to get up before the small group and describe what she had done since the Lovelace testing, Leverton seemed almost transformed with each word. Getting together with the other women had reminded her that there was reason to be proud of having been one of the Mercury 13. After she finished her career update, she thanked the group and confessed that getting together with "her sister pilots" had meant a lot. "It perked me up a little," she said quietly.[3] Eight years later, when Gene Nora Stumbough Jessen invited all the women to meet Eileen Collins, the response was overwhelming. More than thirty years after the Lovelace tests, the Mercury 13 felt they finally had something to celebrate.

Wally Funk, as energetic as she had been at age twenty-two, introduced the Mercury 13 to the crowd of women pilots, family, and friends who had come to the Ninety-Nine International Headquarters near the Oklahoma City Airport for the festive occasion. Wearing a bright red suit and a brilliant smile, Funk called for each woman to stand up as her name was called:

Myrtle Cagle, retired after many years as an airline mechanic in Georgia; Irene Leverton, former women's pylon racing champion and FAA pilot examiner, now living in Arizona; Sarah Gorelick Ratley, a certified public accountant for the federal government in Kansas City; Jerri Sloan Truhill, retired from her company, Air Services, in Dallas and now a grandmother who spent her time volunteering on behalf of mentally retarded children; B Steadman, Powder Puff Derby winner, former president of the Ninety-Nines, and the operator of a taxi service in Traverse City, Michigan; Rhea Hurrle Allison Woltman of Colorado Springs, former seaplane pilot and one of the few registered parliamentarians in the country; Gene Nora Stumbough Jessen, also a former president of the Ninety-Nines and the owner of a Boise, Idaho, flight service; and Wally Funk, former FAA inspector and investigator for the National Transportation and Safety Board, who had taught more than eight hundred men and women how to fly. Funk also spoke in recognition of Janey Hart, who was not able to attend the reunion but sent one of her sons instead. Over the years Hart had developed another passion in addition to flying and political activism and rarely could be pulled away from her sailboat in the Caribbean. Two women did not live to meet the others. Journalist and pilot Marion Dietrich had died of cancer in 1974, and Akron schoolteacher Jean Hixson, who had retired as colonel from the Air Force Reserves, had died of cancer in 1984. Dietrich's twin sister, Jan, who had become a well-respected corporate pilot in California, was seriously ill and no longer able to travel.

As the women looked around at each other, they could not help but feel that the tests, the frustrations, the congressional hearings, and even the final rejection had been worth it. Their efforts had opened the door for women such as Eileen Collins. Looking capable and self-assured, standing ramrod-straight in her Air Force blues and eagerly describing what the upcoming space shuttle mission would accomplish, Collins said, "I know what I want to say, but how can I possibly thank you for what you've done for us?"

Jerrie Cobb was not among those at the banquet. Sometime later she finally appeared, flying from the Amazon, looking worn and tired, with deep lines around her eyes. As she stepped out of her small airplane on the tarmac in Oklahoma City, Cobb looked up as Eileen Collins was the first to greet her. The roar of nearby planes drowned out their private greeting. Wally Funk was standing nearby, overwhelmed with emotion. It had been so many years since Cobb had instructed Funk and Rhea Hurrle to perform sit-ups in her backyard before the isolation tests. The image of the sixty-three-year-old Cobb greeting the woman who would carry all their dreams into space left Funk choking back tears. "It's an emotional time for all of us," she said.[4]

In July 1999, Eileen Collins finally slid into the left seat for her history-making flight. The space shuttle *Columbia*, which four years later disintegrated upon reentry into the earth's atmosphere, would carry Collins and her crew safely into orbit for their mission to deploy the most powerful X-ray telescope ever launched into space. On the ground at Cape Canaveral, bearing witness, were Jerrie Cobb, Janey Hart, Wally Funk, Jerri Sloan Truhill, Sarah Gorelick Ratley, Irene Leverton, B Steadman, and Rhea Hurrle Woltman. Jan Dietrich, in ill health, Myrtle Cagle, and Gene Nora Stumbough Jessen were unable to attend. Collins had invited the surviving Mercury 13 women to be her personal guests at blastoff.* She wanted them to share in the celebration because she genuinely believed the day also belonged to them. "What if they had failed those tests?" Collins asked. It would have reinforced stereotypes and pushed back a women-in-space program even further. Anyone who had passed the tests in 1961 demonstrated that she had the will, the ability, and the courage to go forward, she said, and they all should have been given a chance. Now Collins believed that the time had come to say thank you. Without the Mercury 13, she declared, the country would not be celebrating women astronauts and the first female shuttle commander. "They gave us a history," she said.[5]

The launch of Eileen Collins's mission, STS-93, was delayed several times. In between weather watching and postponements, the eight women who had come for the celebration spent time in Cocoa Beach coffee shops catching up on one another's lives and retelling stories from the past. During breakfast one morning, their group was the only sign of life among sleepy restaurant patrons who huddled silently over toast and coffee. Gales of laughter came from their table as Jerri Sloan Truhill entertained the women with her animated storytelling—tales of kicking off her high heels and flying barefoot across Texas, ditching her crippled plane in a farmer's field, and

*Collins also invited the women who had taken the Lovelace tests but did not pass them. Attending the launch were Georgiana McConnell and Fran Bera. Also attending at Collins's invitation were more than a dozen women who had served as WASPs during World War II.

shocking the old man when a *woman* pilot crawled out of the cockpit. Despite having only one week of their lives in common, the women had become surprisingly close after getting acquainted at reunions. Some traveled together, others telephoned each other once a week. They compared notes on their children and grandchildren, and who among them was still flying and in what kind of aircraft. Now in their sixties and seventies, with some approaching eighty, they counted themselves lucky to be active and healthy and able to attend the historic launch.

Wally Funk announced that she had signed on for a civilian space launch operation, and her fellow pilots understood her ambition, even though some questioned the expense. Interorbital Systems, a business that built rockets and spacecraft at its Mojave Civilian Flight Test Center in California, was opening the door to commercial spaceflight. Funk hoped to be aboard Interorbital's first space mission, to be launched from Tonga, in the South Pacific, sometime in the next five years. Unable to finance a $20 million trip to the orbiting International Space Station, as billionaire Dennis Tito would do in 2001 and as 'N Sync pop star Lance Bass hopes to do, Funk was banking on low-cost private ventures. The Interorbital flight offered the advantage of no governmental red tape from NASA or the Russian space agency and a lower price tag. A deposit of $120,000 was enough to reserve a spot for an "orbital vacation," which was estimated to eventually cost as much as $2 million. Interorbital's flights, however, were still untested, and no one could say with certainty how safe they would be or even exactly when they would be ready to fly. Already Funk's dream of going into space had been delayed several times, but she continued to believe in its future and later spent one night sleeping on the concrete floor of the desert test center just to watch the development of "her rocket." She also had put a deposit down on preliminary astronaut preparation at Star City, Russia, site of the fabled cosmonaut training facility during the Cold War. Strapped for cash, Star City had recently opened its doors to paying customers who wanted a taste of astronaut drills. Now sixty years old, Funk again planned to wear some kind of merry widow for her ride on the old Soviet centrifuge.[6]

Jerrie Cobb joined the others only for brief moments—a cup of coffee, a late-afternoon chat, a lecture on the shuttle's scientific goals. Mainly Cobb kept to herself. No one really knew where she was staying around Cape Canaveral or if she simply spent the nights alone in a car parked along the beach. In the years since they had become better acquainted, the other women had grown used to Cobb's mysteries and her disappearing acts, and they never pressed her for answers. One time, at a large reception, Sarah Gorelick Ratley asked Cobb if she would pose for a snapshot. "It's been so long since we were together," Ratley told Cobb as she went to look for a camera. By the time she returned a few minutes later, Cobb had vanished. Jerrie Cobb has spent too much time in the jungle, some of the women would say; she did not follow the same social customs and behavior as everyone else. But as secretive and remote as Cobb was, the women still respected her. They were impressed after Oklahoma congressman Marvin "Mickey" Edwards nominated Cobb for the Nobel Peace Prize in 1980 for her efforts as a South American missionary pilot. She might be only one woman with one airplane, but she had saved countless lives down there, Truhill noted.[7]

Jerrie Cobb was also on a mission at Cape Canaveral. As ironic as it seemed, John Glenn—unintentionally—had provided her with a renewed opportunity for spaceflight. A year before, seventy-seven-year-old John Glenn had gone into space for a second time, serving as a mission specialist aboard a shuttle flight. NASA said that Glenn had conducted scientific experiments on the effects of aging. Others were not persuaded by NASA's rationale and said that Glenn's return to orbit had delivered a much-needed publicity boost to the flagging space agency. As news of Glenn's second spaceflight began grabbing headlines, J. Donald Dorough, an instructor at Fresno Pacific University, remembered Cobb's struggle. Several years before, he had been assembling a curriculum for a "Women of the West" course and discovered the story of Oklahoma pilot Jerrie Cobb. Dorough wrote a letter to NASA asking why she had not been selected as the senior-citizen-in-space. Supporters, who learned about Dorough's letter through newspaper articles, tracked Cobb down in the Amazon and asked if she would be interested in another shot, if NASA could be persuaded. Cobb was astounded. As difficult as it had been to give up her dream, she knew she would give her life for the chance to be launched into space. "Send Jerrie into Space" soon became a grassroots campaign, with support coming from thousands of school children, the National Organization for Women, women's groups across the country, U.S. senators, and First Lady Hillary Clinton. NASA, however, had not changed its mind.[8] Jerrie Cobb is a remarkable pilot, it said, but we have no plans to launch her into space. That did not stop Cobb from talking with reporters who were interested in her renewed campaign. Nor did it stop her from asking the other women if they would support her effort. As Eileen Collins engaged in last-minute preparations for her historic flight, Cobb corralled the Mercury 13 into a Cocoa Beach motel room and asked them to compose a joint letter to NASA Administrator Daniel Goldin, urging him to give her a chance. Some of the women had doubts that NASA would bring attention to the sexism of 1962, much less redress it. But petitioning the space agency was worth a shot, they believed.

As the hour approached for Eileen Collins's nighttime launch, the women boarded Kennedy Space Center buses and headed to viewing stands along the Banana River across from the launch pad. No one recognized the Mercury 13. They appeared to be just another group of older women with purses, water bottles, and binoculars who had come to Cape Canaveral to watch a shuttle launch. But unlike tourists, who cooled off with Space Dot ice cream or talked on cell phones, Janey Hart stayed put on the metal bleachers and watched the night sky for signs of gathering clouds and lightning. She knew the look of an ominous sky and a storm building from the south, even at night. As the countdown entered its final phase, the crowd started cheering and calling each second: "ten–nine–eight–seven–" Wally Funk's body grew tense with excitement as she whispered to herself, "Go, Eileen. Go for all of us." At six seconds, the clock abruptly froze and mission control called a halt—a technical glitch had been recorded inside the shuttle and the launch was postponed. "T minus six seconds!" a teenager in the stands called out and slammed his fist in frustration on the metal grandstand. "Try T minus thirty-eight years," Truhill muttered under her breath. Then she winked at B Steadman, her Lovelace testing partner.

When the next night came, the women assembled again on the bleachers, a little weary from the endless waiting, the stifling weather, and the insistent mosquitoes. Jerrie Cobb paced alone in front of the large illuminated countdown clock, glancing up only to mark the time or peer at the brilliant lights across the river. Collins said there was nothing in the world like flying the space shuttle; it took off like a rocket, cruised like a spacecraft, and landed like an airplane. Cobb stared at the large external tank cradling the space shuttle *Columbia*, a structure as tall as the Statue of Liberty. She stood still and gazed at a small orange flame that appeared to dance around the base of the launch pad. From the back, she almost looked like twenty-eight-year-old Jerrie Cobb from Ponca City, Oklahoma: a solitary young woman with a lean build, a blond ponytail, and the relaxed posture of an athlete. She never turned around. Staring silently in front of her, Cobb watched the shuttle as steam began to cloud around it and the clock ticked away the seconds. This night the sky was clear. There had been no stalls, and the countdown continued without delay. Cobb kept watching, almost frozen in concentration, until the clock reached five seconds. Then she sat down in the low wet grass. All around her everyone else was standing, cheering, yelling, getting as high in the stands as they could to see the rocket explode off the launch pad. As the roar of engines began and the space shuttle *Columbia* lifted off into the sky, she spread her hands across the grass. Jerrie Cobb wanted to feel the ground rumble.

ACKNOWLEDGMENTS

THIS BOOK WOULD NOT HAVE BEEN POSSIBLE WITHOUT THE HELP OF the Mercury 13. Spending time with these women and their families has been a great personal pleasure, and I want to thank them for sharing so much of their lives with me. Their persistence, resilience, and good humor have been an inspiration.

Of course, the story of the Mercury 13 is also a narrative of physicians, scientists, husbands, daughters, engineers, pilots, friends, politicians, and other individuals who were involved in the early years of the U.S. space program. In researching this complex past, I have benefited greatly from many libraries and archives and want to recognize in particular the NASA History Office, the Dwight D. Eisenhower Presidential Library, the library of *The Daily Oklahoman*, the International Women's Air and Space Museum, the Mount Holyoke College Library, and the headquarters of the Ninety-Nines, the international organization of women pilots. Jane Odom of NASA and the superb library reference staff at Mount Holyoke have been especially helpful in answering my many questions. Aime DeGrenier, of Mount Holyoke, patiently guided me through many technological challenges. I am also grateful to the hundreds of individuals who participated in interviews for this book. Several people merit special recognition for their generosity. They are Ivy Coffey, Pat Daly, Nancy Greep, Jacqueline Lovelace Johnson, Dr. Donald Kilgore, Dr. Cathryn Liberson, Dr. Jack Loeppke, Ruth Lummis, Lawrence Merritt, Dr. Jay T. Shurley, and the late Pauline Vincent.

The Ninety-Nines organization also helped me in ways that reach beyond its archives. I wish to thank Dr. Jacque Boyd and Dr. Petra Illig for helping get this project off the ground and for their useful editorial suggestions. The Ninety-Nines 2001 Amelia Earhart Research Scholars Grant enabled me to travel around the country gathering the oral histories that are central to this book.

My colleagues at Mount Holyoke College have offered me valuable advice and also given me time to concentrate on the work. I would like to extend personal thanks to Christopher Benfey and Karen Remmler for allowing me the mental space to think about outer space and to Jane Crosthwaite, Linda

Laderach, and Sally Montgomery for their good conversation, as always. Kevin McCaffrey's suggestion that I write for an audience beyond the academy came at just the right time and I am deeply grateful for his encouragement. Mount Holyoke alumna Mary McClintock, who has been with this book from the very beginning, gave me the benefit of her astute research ability and her enthusiasm. My appreciation also goes to Dean of Faculty Donal O'Shea, who has supported my work on this book with several faculty grants, the Ellen P. Reese Research award, and a much-needed sabbatical year.

Growing up in St. Louis during the years in which McDonnell Aircraft designed the first space capsule for Project Mercury fueled my early interest in the United States space program. Yet nothing prompted me to look at the sky more than sitting on the hood of our family Chevrolet and watching airplanes take off from Lambert Field. My parents, Florenze and Elizabeth Ackmann, and brothers, David and Rodney, and their families have continued to help me in understanding space and aviation. I appreciate their questions, their suggestions, their legwork, and their hospitality during research trips to St. Louis and Oklahoma.

Lee Boudreaux, my editor at Random House, sharpened the focus of this book during our many useful conversations. Her keen eye and quick intellect have enhanced these pages. I would like to thank my literary agent, Ellen Geiger, whose energy for seeing this story told has been a tonic. My gratitude also goes to Patricia MacLachlan, who was quick to lead me to the indefatigable Ed Wintle of Curtis Brown, Ltd.

Mary Graham Davis deserves special gratitude. Her belief that *The Mercury 13* is much more than a narrative about astronauts has kept me focused on the questions of social equity that this story raises. I thank Mary for her support, her interest in the project, and her friendship.

My deepest thanks go to Ann Romberger, who, over the course of my writing this book, has become far more acquainted with G loads and high-altitude chamber tests than she ever could have imagined. She has been an honest and patient critic. She is in this and all things—to use Emily Dickinson's wonderful words—"a rare ear."

M. A.

LEVERETT, MASSACHUSETTS

NOTES

ABBREVIATIONS

CL	Cathryn Liberson Private Archives
DDE	Jacqueline Cochran Papers, Dwight D. Eisenhower Presidential Library, Abilene, Kansas
GM	Georgiana McConnell Private Archives
HST	James Webb Papers, Harry S. Truman Presidential Library, Independence, Missouri
IWASM	International Women's Air and Space Museum, Cleveland, Ohio
JC	Jerrie Cobb Papers, Ninety-Nines, International Organization of Women Pilots Headquarters, Will Rogers Airport, Oklahoma City, Oklahoma
JS	Dr. Jay T. Shurley Private Archives
JST	Jerri Sloan Truhill Private Archives
LBJ	Vice-Presidential Papers and Presidential Papers, Lyndon Baines Johnson Presidential Library, Austin, Texas
NASA	NASA Historical Reference Collection, National Aeronautics and Space Administration, History Office, NASA Headquarters, Washington, D.C.
NG	Nancy Greep Private Archives
PV	Pauline Vincent Private Archives
SGR	Sarah Gorelick Ratley Private Archives
UNM	University of New Mexico Health Sciences Library and Informatics Center, University of New Mexico, Albuquerque, New Mexico
WF	Wally Funk Private Archives
WP	Wright-Patterson Air Force Base Archives, Dayton, Ohio.

CHAPTER ONE: SPACE FEVER

1. Ivy Coffey, interview with author, January 13, 2002; Jerrie Cobb, *Jerrie Cobb, Solo Pilot* (Sun City Center, Fla.: Jerrie Cobb Foundation, Inc., 1997), 131–139; "Jerrie Cobb Soars to New Record in Aero Commander June 13," *99 News*, July 1957, 4 (JC); Ivy Coffey, "Red Carpet Greets Girl Pilot," *The Daily Oklahoman*, March 21, 1959; Ivy Coffey, "Jerrie's Story," unpublished manuscript (JC); Ruth Lummis, e-mail to author, September 26, 2002.

2. Coffey, "Red Carpet."

3. Ibid.

4. Ibid.

5. Joseph D. Atkinson, Jr., and Jay M. Shafritz, *The Real Stuff: A History of NASA's Astronaut Recruitment Program* (New York: Praeger, 1985), 21.

6. Roger D. Launius, Introduction to *The Birth of NASA: The Diary of T. Keith Glennan* (Washington, D.C.: National Aeronautics and Space Administration, 1993), xix. (Launius cites William E. Burrows, *Deep Black: Space Espionage and National Security* [New York: Random House, 1986], 94.)

7. Atkinson and Shafritz, *The Real Stuff,* 33–37.

8. Chuck Yeager, interview with author, March 23, 2001; John Glenn with Nick Taylor, *John Glenn: A Memoir* (New York: Bantam Books, 2000), 282.

9. "Press Conference Mercury Astronaut Team," April 9, 1959, transcript 3 (NASA).

10. Ibid., 4.

11. "Seven Brave Women Behind the Astronauts," *Life,* September 21, 1959, 142–163; "The Spaceman's Wife: 'Alan Was in His Right Place,' " *Life,* May 12, 1961, 28–29.

12. Jerry Roberts, telephone interview with author, February 1, 2002; Glenn with Taylor, *John Glenn,* 274–275.

13. W. Henry Lambright, *Powering Apollo: James E. Webb of NASA* (Baltimore: Johns Hopkins University Press, 1995), 82.

14. "James Webb Chosen to Head Space Agency, Experienced as Administrator," *The Daily Oklahoman,* January 31, 1961; Lambright, *Powering Apollo,* 77.

15. James Webb, letter to Thomas Harris, February 15, 1961 (JC); Ivy Coffey, e-mail to author, February 14, 2002; "It Takes Snow to Keep Publishers Away," *The Daily Oklahoman,* January 31, 1961; Lambright, *Powering Apollo,* 77; "Kerr Receiving Tribute Friday from Chamber," *The Daily Oklahoman,* January 21, 1961.

16. "From Aviatrix to Astronautrix," *Time,* August 29, 1960, 41; Jerrie Cobb and Jane Rieker, *Woman into Space: The Jerrie Cobb Story* (Englewood Cliffs, N.J.: Prentice Hall, 1963), 155.

17. "12 Women to Take Astronaut Test," *The New York Times,* January 26, 1961.

18. James Webb, oral history interview by T. H. Baker, April 29, 1969 (Internet copy), 6 (LBJ); Lambright, *Powering Apollo,* 84.

19. Webb, oral history, 5.

20. Lambright, *Powering Apollo,* 84–85.

21. "Democrats Praise Kennedy Talk, Call for Bold Challenge," *The Daily Oklahoman,* January 31, 1961; Webb, oral history, 9.

22. "Toward the Endless Frontier," Hearings of the Committee on Science and Technology 1959–79, U.S. House of Representatives (Washington, D.C.: U.S. Government Printing Office, 1980), 80.

23. Mae Mills Link, "Toward Countdown," in *Space Medicine in Project Mercury* (Washington, D.C.: National Aeronautics and Space Administration, Scientific and Technical Information Division, 1965) (Internet copy), 4.

24. "Toward the Endless Frontier," 80–81.

25. "CBS Special Reports," telecast, April 12, 1961, Museum of Television and Radio, New York.

26. Ibid.; Glenn with Taylor, *John Glenn*, 21.

27. Charles Murray and Catherine Bly Cox, *Apollo: The Race to the Moon* (New York: Simon and Schuster, 1989), 80.

28. Joan Fencl Bowski, telephone interview with author, February 18, 2002; Jerry Roberts, telephone interview with author, February 1, 2002; McDonnell Aircraft Project Mercury Engineers Retirement Luncheon participants, interviews with author, January 9, 2002.

29. "Toward the Endless Frontier," 90.

30. Jerrie Cobb, "Woman's Participation in Space Flight," speech to meeting of Aviation/Space Writers Association, New York City, May 1, 1961 (JC).

31. "CBS Special Reports," telecast, May 5, 1961, Museum of Television and Radio, New York.

32. "Toward the Endless Frontier," 89.

33. William E. Burrows, *This New Ocean: The Story of the First Space Age* (New York: Modern Library, 1998), 328–329.

34. Ibid., 329.

35. John F. Kennedy, "Urgent National Needs," telecast speech before joint session of Congress, May 25, 1961, Museum of Television and Radio, New York.

CHAPTER TWO: TAKING A LEAP

1. Ivy Coffey, interview with author, January 13, 2002.

2. Jerrie Cobb, e-mail to author, March 4, 2002.

3. Jerrie Cobb and Jane Rieker, *Woman into Space: The Jerrie Cobb Story* (Englewood Cliffs, N.J.: Prentice Hall, 1963), 18.

4. Carolyn Cobb Lawrence, telephone interview with author, February 19, 2002.

5. Cobb and Rieker, *Woman into Space*, 8.

6. Ibid., 20–21, 9.

7. Ibid., 18–19.

8. Ray Soldan, "Retirement Can't Stop Jim Conger," *Oklahoma City Times*, June 29, 1965; Jerrie Cobb, e-mail to author, March 4, 2002.

9. Cobb and Rieker, *Woman into Space*, 22–25.

10. "1,596 Watch Queens Edge Phoenix, 3–1," *The Daily Oklahoman*, May 10, 1947; Timothy Fisher (Oklahoma City Library System), letter to author, June 18, 2001.

11. "Softball Back, 'N with Curves," *The Daily Oklahoman*, May 10, 1947.

12. Cobb and Rieker, *Woman into Space*, 28.

13. Ibid., 32, 34.

14. Ibid., 28; Ivy Coffey, "The Story of Jerrie Cobb: First American Woman to Qualify as an Astronaut," *Guideposts*, August 1961, 1–5.

15. Joe Evans (Registrar, University of Science and Arts of Oklahoma, formerly Oklahoma College for Women), telephone interview with author, June 14,

2002; Oklahoma College for Women Catalogue, 1949, 18–19; Cobb and Rieker, *Woman into Space*, 29.

16. Oklahoma College for Women Catalogue, 19.

17. Cobb and Rieker, *Woman into Space*, 30–31; Jerrie Cobb, e-mail to author, March 4, 2002.

18. Cobb and Rieker, *Woman into Space*, 47.

19. Susie Sewell, telephone interview with author, October 15, 2002.

20. Sarah Gorelick Ratley, interview with author, January 12, 2002.

21. Jerri Sloan Truhill, interview with author, July 20, 2001; Jerrie Cobb, e-mail to author, March 4, 2002.

22. Cobb and Rieker, *Woman into Space*, 47–57.

23. Ibid., 61–62.

24. Ibid., 61–83.

25. Ibid., 75.

26. Ibid., 74.

27. Ibid., 24.

28. Ibid., 14.

29. Ibid., 109.

30. Ibid.

31. Ibid., 111, quoting *Flight* magazine.

32. Ivy Coffey, "Jerrie's Story," unpublished manuscript (JC).

33. Ivy Coffey, interview with author, January 13, 2002.

34. Ivy Coffey, "Red Carpet Greets Girl Pilot," *The Daily Oklahoman*, March 21, 1959; Cobb and Rieker, *Woman into Space*, 128–130.

35. Bernice Steadman, interview with author, September 26, 2001.

36. Frank Sis, "Say Astronauts to Get Company," *The Cleveland Press*, July 18, 1961.

37. Chuck Yeager and Leo Janos, *Yeager* (New York: Bantam, 1985), 272.

38. Jacqueline Cochran, *The Stars at Noon* (Boston: Little, Brown, 1954), 429.

39. Yeager and Janos, *Yeager*, 275.

40. Chuck Yeager, interview with author, March 23, 2001.

41. Jacqueline Cochran and Maryann Bucknum Brinley, *Jacqueline Cochran: An Autobiography* (New York: Bantam Books, 1987), 211; Harold Newcomb, "Cochran's Convent," *Airman*, May 20, 1977, n.p. (NASA); Leslie Haynsworth and David Toomey, *Amelia Earhart's Daughters: The Wild and Glorious Story of American Women Aviators from World War II to the Dawn of the Space Age* (New York: Morrow, 1998), 97, 124–125, 129–132, 141–144; Julie I. Englund, "First-Rate, Second-Class," *The Washington Post*, May 13, 2002; Dr. Jacque Boyd, interview with author, Reno, Nevada, March 25, 2001.

42. Jacqueline Lovelace Johnson, interview with author, January 11, 2002.

43. Robert Secrest, oral history interview by Jake Spidle, July 8, 1996, Oral History of Medicine Project (UNM), 10.

44. Margaret Weitekamp, "The Right Stuff, The Wrong Sex: The Science,

Culture, and Politics of the Lovelace Woman in Space Program, 1959–1963," Ph.D. dissertation, Cornell University, May 2001, 50.

45. Richard G. Elliott, " 'On a Comet, Always': A Biography of W. Randolph Lovelace II," *New Mexico Quarterly* 36 (1966–1967), 361–362 (UNM).

46. "Army Doctor's Record Parachute Jump," *Life*, August 9, 1943, 69.

47. Elliott, " 'On a Comet, Always,' " 364; Edward T. Martin, "The Hero at High Altitude Flight," *Airline Pilot*, February 1983, 22–24, 38; "Leap from the Stratosphere," *Boeing News*, August 1943, 3–4, 6.

48. Shirley Thomas, "William Randolph Lovelace II," *Men of Space: Profiles of the Leaders in Space Research, Development, and Exploration*, vol. 2 (Philadelphia: Chilton Company, 1961), 98.

49. Elliott, " 'On a Comet, Always,' " 363; Martin, "The Hero at High Altitude Flight," 38.

50. Joe Godfrey, NASA Biographical Data Fact Sheet and audiovisual Web profile.

51. Scott Crossfield, telephone interview with author, April 1, 2002.

52. Donald E. Kilgore, oral history interview by Jake Spidle, November 18 and 25 and December 9, 1985, Oral History of Medicine Project (UNM), 3.

53. Donald Kilgore, interview with author, August 10, 2000.

54. W. Randolph Lovelace II, "Human Factors in Space Exploration," unpublished manuscript (UNM).

55. Cobb and Rieker, *Woman into Space*, 130–133.

CHAPTER THREE: GIRL ASTRONAUT PROGRAM

1. Stanley Mohler, telephone interview with author, April 17, 2002; Stanley Mohler, letter to author, April 22, 2002.

2. Stanley Mohler, telephone interview with author, April 17, 2002.

3. "Dr. Donald D. Flickinger, 89, A Pioneer in Space Medicine," *The New York Times*, March 3, 1997; "NASA Press Conference," April 9, 1959, transcript 23.

4. Shirley Thomas, "Don D. Flickinger: With Zest and Dedication, This Energetic Doctor Has Long Concentrated on the Problems of Man's Survival in the Hostile Environment of Space," *Men of Space: Profiles of the Leaders in Space Research, Development, and Exploration*, vol. 3 (Philadelphia: Chilton Company, 1961), 75.

5. Walter Bonney, NASA spokesman, "NASA Press Conference," April 9, 1959, transcript 15.

6. Don Flickinger, "Action Memorandum" to W. Randolph Lovelace II, December 20, 1959 (JC).

7. Jerri Sloan Truhill, e-mail to author, March 19, 2002.

8. Daphne Flickinger Bradford, telephone interview with author, October 19, 2002.

9. Stanley Mohler, telephone interview with author, April 17, 2002.

10. Warren Young, "What It's Like to Fly in Space," *Life,* April 13, 1959, 133–148.

11. "Reminiscences of Ruth Nichols," interview by Kenneth Leish, June 1960, in the Oral History Collection, Part IV (1–219), Columbia University, New York City, 1.

12. "Ruth Nichols Rivals Amelia Earhart in Aviation Accomplishments," *Foundation for the Carolinas Newsletter,* Winter 1999.

13. Ruth Nichols, "Reminiscences," 12.

14. Ibid., 26.

15. Ruth Nichols, *Wings for Life* (Philadelphia: Lippincott, 1957), 144.

16. Nichols, "Reminiscences," 27.

17. Nichols, *Wings for Life,* 204.

18. Henry Holden and Lori Griffith, *Ladybirds II: The Continuing Story of American Women in Aviation* (Mt. Freedom, N.J.: Blackhawk Publishing Co., 1993), 29.

19. Nichols, "Reminiscences," 32–33.

20. Ruth Nichols, "Why Not Lady Astronauts?," *Washington Daily News,* November 24, 1959.

21. "First Jet Ride," *Boston Traveler,* July 7, 1955.

22. Tom Renner, "Aviatrix Flies 1,000 MPH over LI, Sets New Mark," *Newsday,* January 22, 1958.

23. Carol Gelber, "Ruth Nichols Flies Through the Years," *Philadelphia Evening Bulletin,* May 28, 1959; "Jet Record Claimed," *The New York Times,* January 22, 1958.

24. "Women Best Suited for Space, Pioneer Aviatrix Says," *The Washington Post,* April 16, 1960; "Reminiscences," 39.

25. Nichols, "Reminiscences," 39–40.

26. Ibid., 42.

27. Ibid.

28. Nichols, "Why Not Lady Astronauts?"

29. Dorothy Roe, "Space Is Goal of Gal Flyer," *The Baltimore Evening Sun,* February 14, 1958.

30. Nichols, "Reminiscences," 33–34.

31. Nichols, *Wings for Life,* 293.

32. Nichols, "Why Not Lady Astronauts?"

33. Nichols, "Reminiscences," 40–42.

34. Ibid., 42–43.

35. Ibid.

36. Ibid., 43.

37. *"The Star,"* April 14, 1959 (NASA).

38. Nichols, "Why Not Lady Astronauts?"

39. Don Flickinger, "Action Memorandum" to W. Randolph Lovelace II, December 20, 1959 (JC).

40. Don Flickinger, letter to Jerrie Cobb, December 7, 1959 (JC).

41. Ibid.

42. Don Flickinger, "Action Memorandum" to W. Randolph Lovelace II, December 20, 1959 (JC).

43. Stanley Mohler, telephone interview with author, April 17, 2002.

44. Don Flickinger, "Action Memorandum" to W. Randolph Lovelace II, December 20, 1959 (JC).

45. Dryden Flight Research Center, "A Brief History of the Pressure Suit" (www.dfrc.nasa.gov/airsci/er-2/pshis.html); Petra Illig, e-mail to author, November 5, 2002.

46. www.sothebys.com/live/auctions/sneak/archives/minispace/lot19.html.

47. Scott Crossfield, telephone interview with author, April 1, 2002.

48. Ibid.

49. Don Flickinger, "Action Memorandum" to W. Randolph Lovelace II, December 20, 1959 (JC).

50. Ibid.

51. Ibid.

52. William E. Burrows, *This New Ocean: The Story of the First Space Age* (New York: Modern Library, 1998), 262–263.

53. Ibid., 259.

54. Donald Kilgore, interview with author, August 10, 2000.

55. Burrows, *This New Ocean*, 266.

CHAPTER FOUR:
THE VIEW FROM ALBUQUERQUE

1. Ivy Coffey, "Jerrie Cobb Flies Jet Fighter: Pilot Adds Another 'First' to Career," *The Daily Oklahoman*, October 29, 1959.

2. Ivy Coffey, interview with author, January 13, 2002.

3. Coffey, "Jerrie Cobb Flies Jet Fighter."

4. Ibid.

5. Ann Gennett, "Space for Women Too," *Contributions of Women: Aviation* (Minneapolis: Dillon Press, 1975); Coffey, "Jerrie Cobb Flies Jet Fighter."

6. Chuck Yeager, interview with author, March 23, 2001.

7. Chuck Yeager and Leo Janos, *Yeager* (New York: Bantam, 1985), 282–283.

8. Ibid., 285; Jacqueline Cochran and Maryann Bucknum Brinley, *Jacqueline Cochran: An Autobiography* (New York: Bantam Books, 1987), 277.

9. Cochran and Brinley, *Jacqueline Cochran*, 276.

10. Ivy Coffey, interview with author, January 13, 2002; Ivy Coffey, "Meet Jerrie Cobb—First Woman-in-Space Candidate," *The Daily Oklahoman*, n.d. (JC).

11. Ivy Coffey, "First Spacelady," *The Daily Oklahoman*, October 23, 1960; Jerrie Cobb and Jane Rieker, *Woman into Space: The Jerrie Cobb Story* (Englewood Cliffs, N.J.: Prentice Hall, 1963), 135–136.

12. Coffey, "Meet Jerrie Cobb."

13. Randy Lovelace, "Duckings, Probings, Checks That Proved Fliers' Fitness," *Life*, April 20, 1959, 26.

14. Ibid., 26–27.

15. Ibid., 26.

16. Ibid.

17. Ibid.

18. Richard G. Elliott, " 'On a Comet, Always': A Biography of W. Randolph Lovelace II," *New Mexico Quarterly* 36 (1966–1967), 357–358 (UNM).

19. Ben Kocivar, producer, "The Lady Wants to Orbit," *Look*, February 2, 1960, 113.

20. Betty Skelton Frankman, interview by Carol L. Butler, July 19, 1999, Cocoa Beach, Florida, NASA Oral History Project, 9.

21. Ibid., 11.

22. Kocivar, "The Lady Wants to Orbit," 114.

23. Frankman, oral history, 18.

24. Ibid., 33; Kocivar, "The Lady Wants to Orbit," 112.

25. Frankman, oral history, 17.

26. Ibid., 40.

27. Kocivar, "The Lady Wants to Orbit," 117.

28. Ibid., 116.

29. Jake W. Spidle, Jr., *The Lovelace Medical Center: Toward the 21st Century* (Albuquerque: University of New Mexico Press, 1992), 2.

30. Jake W. Spidle, Jr., *The Lovelace Medical Center: Pioneer in American Health Care* (Albuquerque: University of New Mexico Press, 1987), 1–90; "Lovelace Foundation for Medical Education and Research," *Albuquerque Journal*, n.d. (UNM); Jake W. Spidle, Jr., interview with author, August 9, 2000.

31. Donald E. Kilgore, oral history interview by Jake Spidle, Jr., November 18 and 25 and December 9, 1985, Oral History of Medicine Project, 18 (UNM).

32. Mercury Astronaut Selection Fact Sheet, April 9, 1959 (NASA).

33. Robert Secrest, oral history interview by Jake Spidle, Jr., July 8, 1996, Oral History of Medicine Project, 12 (UNM).

34. W. Randolph Lovelace II, A. H. Schwichtenberg, Ulrich C. Luft, and Robert R. Secrest, "Selection and Maintenance Program for Astronauts for the National Aeronautics and Space Administration," *Aerospace Medicine*, June 1962, 667–684.

35. Cobb and Rieker, *Woman into Space*, 139–140; "A Lady Proves She's Fit for Space Flight," *Life*, August 29, 1960, 72.

36. NASA, press release, April 9, 1959.

37. Cobb and Rieker, *Woman into Space*, 141.

38. Jacqueline Lovelace Johnson, interview with author, January 11, 2002.

39. "Under Sheltering Wings: Secluded Siesta Hills Area Enjoys Sounds of Air Traffic," *Albuquerque Journal*, December 28, 1991.

40. Cobb and Rieker, *Woman into Space*, 144.

41. Ibid.

42. A. H. Schwichtenberg, oral history interview by Jake Spidle Jr., February 20, 1985, Oral History of Medicine Project, 12 (UNM).

43. Ibid., 11.

44. Cobb and Rieker, *Woman into Space*, 147.

45. Lovelace, "Duckings, Probings, Checks That Proved Fliers' Fitness," 26.

46. Cobb and Rieker, *Woman into Space*, 143.

47. Lovelace, "Duckings, Probings, Checks That Proved Fliers' Fitness," 26.

48. Ray Ward Taylor, letter to Jerrie Cobb, February 20, 1961 (JC).

49. John Glenn with Nick Taylor, *John Glenn: A Memoir* (New York: Bantam Books, 2000), 288.

50. Ibid., 289.

51. Cobb and Rieker, *Woman into Space*, 152.

52. W. Randolph Lovelace II, letter to Jane Rieker, n.d. (JC).

53. Cobb and Rieker, *Woman into Space*, 155.

54. Ibid., 156.

55. "A Woman Passes Tests Given to 7 Astronauts," *The New York Times*, August 19, 1960.

56. Ivy Coffey, "City Woman's Eager to Make Pioneering Flight into Space," *The Daily Oklahoman*, August 19, 1960.

57. Jane Rieker, "Up and Up Goes Jerrie Cobb," *Sports Illustrated*, August 29, 1960, 28–29.

58. Cobb and Rieker, *Woman into Space*, 158.

59. Sidney Fields, "The Indestructible First Lady of World Aviation," *The New York Mirror*, January 12, 1958.

60. "Death Ruled a Suicide," *The New York Times*, October 20, 1960; "Police Suspect Suicide: Ruth R. Nichols, Famed Flier Dies in Apartment," *The Boston Globe*, September 26, 1960; "Ruth Nichols Death Here Listed by Police as Possible Suicide," *The New York Times*, September 26, 1960; "Ruth Nichols Found Dead, May Have Been a Suicide," *Boston Herald Tribune*, September 27, 1960.

61. Jane Hyde Fawcett, letter to Alice Benson, October 11, 1960; Alice Benson, letter to Jane Hyde Fawcett, October 20, 1960 (NG).

CHAPTER FIVE: THE LIST

1. Dr. Jacque Boyd, e-mail to author, October 16, 2002.

2. Jerrie Cobb, e-mail to author, June 10, 2002; Jerrie Cobb and Jane Rieker, *Woman into Space: The Jerrie Cobb Story* (Englewood Cliffs, N.J.: Prentice Hall, 1963), 203–204.

3. Sarah Gorelick Ratley, interview with author, January 12, 2002; Wally Funk, telephone interview with author, June 14, 2002; "Qualifications for Astronauts. Hearings before the Special Subcommittee on the Selection of Astronauts," Committee on Science and Astronautics, U.S. House of Representatives, July 17 and 18, 1962 (Washington, D.C.: U.S. Government Printing Office, 1962), 4–5; Jacqueline Cochran, letters to W. Randolph Lovelace II, November 28, 1960, and January 31, 1961 (DDE).

4. Jacqueline Cochran, interoffice memorandum, December 6, 1960 (DDE).
5. Jerrie Cobb, e-mail to author, June 10, 2002; Rufus Hunt, telephone interview with author, December 2, 2002.
6. Jerri Sloan Truhill, interview with author, July 20, 2001; Jerri Sloan Truhill, private archives.
7. Jerri Sloan Truhill, interview with author, July 20, 2001; Jerri Sloan Truhill, private archives; Jerri Sloan Truhill, telephone interview with author, May 14, 2002.
8. Marion Dietrich, letter to Jan Dietrich, January 13, 1961 (IWASM).
9. Marion Dietrich, "First Woman into Space," *McCall's*, September 1961, 180.
10. Wally Funk, interview with author, December 4, 1998; Wally Funk, private archives.
11. Gene Nora Stumbough Jessen, telephone interview with author, June 10, 2002.
12. Bernice Steadman, interview with author, September 26, 2001; Jane Hart, interview with author, September 26, 2001.
13. Betty Marsh, letter to Jean Hixson, n.d. (PV).
14. Helen Waterhouse, "Ohio Teacher Crashes Sound Barrier in Jet," *The Christian Science Monitor,* March 21, 1957; Pauline Vincent, interview with author, June 17, 2002; Jeanne Randles, telephone interview with author, June 14, 2001.
15. Pauline Vincent, private archives; Pauline Vincent, interview with author, June 17, 2002.
16. "Girls Ride into Space Still Long Way Off," *The Dayton Daily News*, September 29, 1960.
17. Donald Cox, "Woman Astronauts," *Space World*, September 1961, 59.
18. "Updates to Jackie Cochran," n.d. (DDE); "List of Women Who Completed Medical Tests at Lovelace Foundation, Albuquerque, N.M." n.d. (DDE); Dorothy Anderson, telephone interview with author, June 1, 2002; Frances Bera, telephone interview with author, May 14, 2002; Patricia Jetton, telephone interview with author, May 23, 2002; Georgiana McConnell, telephone interview with author, May 13, 2002; Sylvia Roth and Frances Miller, telephone interview with author, May 24, 2002; W. R. Lovelace II, letter to Georgiana McConnell, May 22, 1961 (GM); Georgiana McConnell, "Written after returning from Albuquerque in 1961," n.d. (GM).
19. Jacqueline Cochran and Maryann Bucknum Brinley, *Jacqueline Cochran: An Autobiography* (New York: Bantam Books, 1987), 29.
20. Jacqueline Cochran, letter to W. R. Lovelace II, November 28, 1960 (DDE).
21. Floyd Odlum to W. R. Lovelace II, February 6, 1961, and November 17, 1961 (DDE).
22. Jacqueline Cochran, letter to W. R. Lovelace II, December 27, 1960 (DDE).
23. Ibid.
24. Ruby Clayton McKee, "Jacqueline Cochran Stresses Top Role for Women in Space," *The Dallas Morning News*, January 18, 1961; Jerri Sloan Truhill,

interview with author, July 20, 2001; Patricia Jetton, telephone interview with author, May 23, 2002.

25.   Chuck Yeager and Leo Janos, *Yeager* (New York: Bantam, 1985), 276.

26.   Jerri Sloan Truhill, interview with author, July 20, 2001; Jerri Sloan Truhill, telephone interview with author, May 14, 2002.

CHAPTER SIX: THE BIRD OF PARADISE

1.   Jacqueline Lovelace Johnson, interview with author, January 11, 2002.

2.   Jerrie Cobb, letter to Jacqueline Cochran, February 20, 1961; Jacqueline Cochran, telegram to Jerrie Cobb, June 18, 1961; Jerrie Cobb, telegram to Jacqueline Cochran, June 20, 1961; Jerrie Cobb, letter to Jacqueline Cochran, June 14, 1961; Jacqueline Cochran, secretary's note, May 31, 1961; Jerrie Cobb, postcard to Jacqueline Cochran, February 14, 1961; Jacqueline Cochran, letter to Jerrie Cobb, January 30, 1961; Jerrie Cobb, telegram to Jacqueline Cochran, January 18, 1961 (all DDE).

3.   Sarah Gorelick Ratley, interview with author, January 12, 2002.

4.   Patricia Jetton, telephone interview with author, May 23, 2002.

5.   Donald Kilgore, interview with author, August 10, 2000; Sarah Gorelick Ratley, interview with author, January 12, 2002; Jerri Sloan Truhill, interview with author, July 20, 2001.

6.   Jan Dietrich, letter to Jackie Cochran, February 21, 1961 (DDE); Jan Dietrich, letter to Floyd Odlum, January 23, 1961 (DDE).

7.   Marion Dietrich, "First Woman into Space," *McCall's*, September 1961, 180.

8.   Jan Dietrich, letter to Floyd Odlum, January 23, 1961 (DDE).

9.   Jeanne Williams, letter to Jacqueline Cochran, November 28, 1960 (DDE).

10.   Dietrich, "First Woman into Space," 180, 182.

11.   Marion Dietrich, letter to Jan Dietrich, January 13, 1961 (IWASM).

12.   Jacqueline Cochran, "Women in Space: Famed Aviatrix Predicts Women Astronauts Within Six Years," *Parade*, April 30, 1961, 8.

13.   Marion Dietrich, letter to Jacqueline Cochran, May 16, 1961 (DDE).

14.   Jacqueline Cochran, letter to Marion Dietrich, July 12, 1961 (DDE).

15.   Edith Hills Coogler, "Possible Woman Astronaut: Tar Heel Set Sights on Moon," *Charlotte Observer*, July 14, 1963; Myrtle Cagle, letters to Jacqueline Cochran, May 1, 1961, June 20, 1961, and September 5, 1961 (DDE); "Qualifications for Astronauts. Hearings before the Special Subcommittee on the Selection of Astronauts," Committee on Science and Astronautics, U.S. House of Representatives, July 17 and 18, 1962 (Washington, D.C.: U.S. Government Printing Office, 1962).

16.   Sarah Gorelick Ratley, interview with author, January 12, 2002.

17.   Jerri Sloan Truhill, interview with author, July 20, 2001.

18.   Ibid.; Bernice Steadman, interview with author, September 26, 2001.

19.   Ibid.

20.   Jerri Sloan Truhill, interview with author, July 20, 2001.

21. Wally Funk, interview with author, December 4, 1998; Donald E. Kilgore, interview with author, August 10, 2000.

22. Cochran, "Women in Space," 8.

23. Gene Nora Stumbough Jessen, telephone interview with author, June 10, 2002; Donald E. Kilgore, interview with author, August 10, 2000.

24. Jane Hart, interviews with author, September 26 and November 26, 2001.

25. Donald E. Kilgore, interview with author, August 10, 2000; Gene Nora Stumbough Jessen, telephone interview with author, June 10, 2002; Jane Hart, interview with author, November 26, 2001.

26. Margaret Weitekamp, "The Right Stuff, The Wrong Sex: The Science, Culture, and Politics of the Lovelace Woman in Space Program, 1959–1963," Ph.D. dissertation, Cornell University, May 2001, 235–236.

27. Floyd Odlum, letter to Jacqueline Cochran, March 6, 1961 (DDE).

28. Weitekamp, "The Right Stuff," 236.

29. Frances Bera, telephone interview with author, May 14, 2002; Patricia Jetton, telephone interview with author, May 23, 2002.

30. Betty J. Miller, telephone interview with author, May 23, 2002.

31. Rhea Hurrle Allison Woltman, interview with author, July 16, 2002.

32. Irene Leverton, interview with author, July 18, 2001.

33. "Like Man: Women Can Be 'Way Out,' Too," *The Washington Post*, May 21, 1961.

34. Donald E. Kilgore, interview with author, August 10, 2000; Bernice Steadman, interview with author, September 26, 2001.

35. Irene Leverton, interview with author, July 18, 2001.

CHAPTER SEVEN: PROJECT VENUS

1. W. Randolph Lovelace II, letter to Wally Funk, May 17, 1961 (WF); W. Randolph Lovelace II, letter to Jerrie Cobb, May 17, 1961 (JC).

2. Jay T. Shurley, interview with author, January 15, 2002; Jay T. Shurley, telephone interview with author, May 17, 2002.

3. Richard Green, "The Early Years: Jolly West and the University of Oklahoma Department of Psychiatry," *Journal of the Oklahoma State Medical Association* 93, no. 9 (September 2000), 446–454; Jay T. Shurley, interview with author, January 15, 2002.

4. Green, "The Early Years," 449.

5. Jay T. Shurley, edited copy of NASA Press Release no. 59-113, "Mercury Astronaut Selection Fact Sheet" (JS).

6. Jay T. Shurley, interview with author, January 15, 2002.

7. Ibid.

8. Jay T. Shurley, telephone interview with author, November 12, 2001.

9. Jay T. Shurley, telephone interview with author, May 17, 2002.

10. George E. Ruff and Edwin Z. Levy, "Psychiatric Evaluation of Candidates for Space Flight," *The Journal of Psychiatry* 116, no. 5 (November 1959), 388.

11. Ibid., 390.

12. John Glenn with Nick Taylor, *John Glenn: A Memoir* (New York: Bantam Books, 2000), 249–251.

13. Jay T. Shurley, "Mental Images in Profound Experimental Sensory Isolation," in *Hallucinations*, edited by Louis Jolyon West (New York: Grune & Stratton, 1962), 154–156.

14. Covey Bean, "OU Professor Misses Out on Profits of Latest Fad," *The Daily Oklahoman*, October 15, 1979.

15. "The Tank," videotape, WKY-TV, Oklahoma City, 1960 (JC).

16. Jay T. Shurley, "The Hydro-Hypodynamic Environment," in *Proceedings of the Third World Congress of Psychiatry*, vol. 3 (Montreal: McGill University Press, 1961), 235.

17. Cathryn (Walters) Liberson, telephone interview with author, June 29, 2002.

18. Shurley, "Mental Images in Profound Experimental Sensory Isolation," 539.

19. Ibid., 543.

20. Jay T. Shurley, interview with author, January 15, 2002.

21. Ibid.

22. Ibid.

23. Shurley, "The Hydro-Hypodynamic Environment," 236, quoting Freud's *Interpretation of Dreams*.

24. John C. Lilly and Jay T. Shurley, "Experiments in Solitude, in Maximum Achievable Physical Isolation with Water Suspension, of Intact Healthy Persons," in *Psychophysiological Aspects of Space Flight*, edited by Bernard E. Flaherty (New York: Columbia University Press, 1961), 245–246.

25. Shurley, "Mental Images in Profound Experimental Sensory Isolation," 541–542.

26. Jay T. Shurley, interview with author, January 15, 2002.

27. Ibid.

28. Shurley, "The Hydro-Hypodynamic Environment," 235.

29. "The Tank," videotape, WKY-TV, Oklahoma City, 1960 (JS).

30. Jay T. Shurley, interview with author, January 15, 2002.

31. Ibid.; "Damp Prelude to Space: A Potential Lady Orbiter Excels in Lonesome Test," *Life*, October 24, 1960, 81; Jay T. Shurley and Cathryn Walters, "Woman Astronaut Assessment in Hydrohypodynamic Environment," August 8, 1961 (JS, JC).

32. Jerrie Cobb and Jane Rieker, *Woman into Space: The Jerrie Cobb Story* (Englewood Cliffs, N.J.: Prentice Hall, 1963), 167, 172–173; Shurley and Walters, "Woman Astronaut Assessment."

33. Jay T. Shurley, interview with author, January 15, 2002; Shurley and Walters, "Woman Astronaut Assessment."

34. Jay T. Shurley, interview with author, January 15, 2002.

35. Shurley and Walters, "Woman Astronaut Assessment"; "Qualifications for Astronauts. Hearings before the Special Subcommittee on the Selection of

Astronauts," Committee on Science and Astronautics, U.S. House of Representatives, July 17 and 18, 1962 (Washington, D.C.: U.S. Government Printing Office, 1962), 83.

36. Ivy Coffey, "The Story of Jerrie Cobb: First American Woman to Qualify as an Astronaut," *Guideposts*, August 1961, 2.

37. Jerri Sloan Truhill, interview with author, July 20, 2001; Jerrie Cobb, letter to F.L.A.T.s, July 24, 1961 (IWASM).

38. Cobb and Rieker, *Woman into Space*, 204.

39. Ibid., 205; Wally Funk, interview with author, December 4, 1998; Rhea Hurrle Allison Woltman, interview with author, July 16, 2002.

40. Shurley and Walters, "Woman Astronaut Assessment"; Jay T. Shurley, Project Venus files (JS).

41. Jay T. Shurley, "Cost Ceiling Estimate" (JS); Jay T. Shurley, Project Venus files (JS).

CHAPTER EIGHT: WAITING FOR PENSACOLA

1. Burt Boldt, telephone interview with author, February 14, 2002; Jane Hoffstetter, "She'd Be First Woman in Space," *Fort Lauderdale News*, May 19, 1961.

2. Jerrie Cobb and Jane Rieker, *Woman into Space: The Jerrie Cobb Story* (Englewood Cliffs, N.J.: Prentice Hall, 1963), 196.

3. Ibid., 94–95.

4. Ibid., 196–197.

5. Ibid., 198.

6. Ibid., 199–201.

7. Jerrie Cobb, e-mail to author, August 10, 2002.

8. W. Randolph Lovelace II, letter to Wally Funk, July 8, 1961 (WF).

9. Jerrie Cobb, letter to James Webb, May 16, 1961 (JC).

10. Chuck Wheat, "Governor Late, Kerr Tired—But Space Meet Launched," *Tulsa World*, May 26, 1961; "DWD Jr., Other Top Douglas Officials Will Be Here to Take Part in World Space Conference," *Tulsa World*, Peaceful Uses of Space Section, May 27, 1961; "Experts on Space Research to Appear Here," *Tulsa World*, Peaceful Uses of Space Section, May 27, 1961; "Space Future for Women: Test Foundation to Be Expanded," *Tulsa World*, May 27, 1961; "Four States Bid for '62 National Space Parlay," *Tulsa World*, May 28, 1961.

11. NASA, memorandum, "Proposed Answers to Female Astronaut Volunteers," May 31, 1961 (NASA).

12. Jerrie Cobb, "Women as Astronauts," unpublished manuscript (JC).

13. Jerrie Cobb, letter to James Webb, May 30, 1960 (JC).

14. Jerrie Cobb, letter to Jacqueline Cochran, May 31, 1961 (DDE).

15. Floyd Odlum, letter to W. Randolph Lovelace II, May 31, 1961 (DDE).

16. W. Randolph Lovelace II, letter to Floyd Odlum, June 8, 1961 (DDE).

17. Ibid.

18. Floyd Odlum, letter to W. Randolph Lovelace II, May 31, 1961 (DDE).

19. Jerrie Cobb, letter to F.L.A.T.s, May 29, 1961 (JST); Jerri Sloan, release form (IWASM).

20. W. Randolph Lovelace II, letter to Jerri Sloan, July 12, 1961 (JST).

21. Jacqueline Cochran, letter to Marion Dietrich, July 12, 1961 (DDE).

22. Sarah Gorelick Ratley, interview with author, January 12, 2002; Margaret Weitekamp, "The Right Stuff, The Wrong Sex: The Science, Culture, and Politics of the Lovelace Woman in Space Program, 1959–1963," Ph.D. dissertation, Cornell University, May 2001, 248.

23. Jerrie Cobb, Report to NASA, June 15, 1961 (NASA).

24. Jacqueline Cochran, "personal and confidential" memorandum to Admiral Robert Pirie, August 1, 1961 (DDE).

25. "Qualifications for Astronauts. Hearings before the Special Subcommittee on the Selection of Astronauts," Committee on Science and Astronautics, U.S. House of Representatives, July 17 and 18, 1962 (Washington, D.C.: U.S. Government Printing Office, 1962), 11.

26. Joseph D. Atkinson, Jr., and Jay M. Shafritz, *The Real Stuff: A History of NASA's Astronaut Recruitment Program* (New York: Praeger, 1985), 90.

27. Northrop, press release, October 17, 1961 (DDE).

28. Jacqueline Cochran, letter to W. Randolph Lovelace II, September 1, 1961 (DDE).

29. Chuck Yeager, diary, October 12, 1961 (DDE); Jaqueline Cochran and Maryann Bucknum Brinley, *Jaqueline Cochran: An Autobiography* (New York: Bantam Books, 1987), 308.

30. Sarah Gorelick Ratley, interview with author, January 12, 2002.

31. Lovelace Foundation, telegram to Wally Funk, September 12, 1961 (WF).

32. Hugh Dryden, letter to Vice Admiral R. B. Pirie, October 2, 1961 (NASA).

33. Jacqueline Cochran, memo to secretary, October 17, 1961 (DDE).

CHAPTER NINE: CHANGING COURSE

1. Liz Carpenter, memorandum to the Vice President, "Background for Your Conference at 11 a.m. Thursday on Women in Space," March 14, 1962 (LBJ).

2. W. Randolph Lovelace II, letter to James Webb, September 29, 1961 (DDE).

3. Edward R. Murrow, letter to James Webb, September 21, 1961 (NASA).

4. Joseph D. Atkinson, Jr., and Jay M. Shafritz, *The Real Stuff: A History of NASA's Astronaut Recruitment Program* (New York: Praeger, 1985), 98–100, quoting Dr. Robert B. Voas, Human Factors Assistant to the Director of the Manned Space Center.

5. Jerrie Cobb, e-mail to author, July 20, 2002.

6. "Cobb's speaking" Bonnie [?] and Jerrie Cobb to Jane [?], December 7, 1961 (JC).

7. Hugh Dryden, letter to Shirley Thomas, December 8, 1961 (NASA).

8. James Webb, letter to Jerrie Cobb, December 15, 1961 (JC).

9. Hugh Dryden, letter to Shirley Thomas, December 20, 1961 (NASA).

10. Jerrie Cobb, e-mail to author, July 20, 2002.

11. Jerrie Cobb, letter to F.L.A.T.s, February 12, 1962 (IWASM).

12. Ibid.

13. Jerrie Cobb, memorandum to Lyndon Johnson, "Space for Women," April 17, 1962 (LBJ); "First Woman Astronaut Trainee Tells of Program for Space," *Los Angeles Times*, February 23, 1962; Jerrie Cobb, letter to Lyndon Johnson, April 17, 1962 (LBJ).

14. Shirley Thomas, telephone interview with author, April 2, 2002.

15. Jacqueline Cochran, letter to W. Randolph Lovelace II, February 27, 1962 (DDE).

16. John Glenn with Nick Taylor, *John Glenn: A Memoir* (New York: Bantam Books, 2000), 366.

17. Floyd Odlum, letter to W. Randolph Lovelace II, March 29, 1962 (DDE).

18. Ivy Coffey, interview with author, January 13, 2002; Jay T. Shurley, interview with author, January 15, 2002.

19. Jacqueline Cochran, letter to Jerrie Cobb, March 23, 1962 (DDE).

20. Ibid.

21. Jerrie Cobb, letter to Jacqueline Cochran, April 26, 1962 (DDE).

22. Sarah Gorelick Ratley, interview with author, January 12, 2002.

23. Gene Nora Stumbough, letter to Jacqueline Cochran, June 11, 1962 (DDE).

24. Jane Hart, interviews with author, September 26 and November 26, 2001.

25. Jane Hart, interview with author, September 26, 2001.

26. Carpenter, memorandum to the Vice President.

27. "Woman-in-Space Program Urged," *The Washington Post*, n.d. (DDE).

28. Ibid.

29. Glenn with Taylor, *John Glenn*, 369, 376.

30. Carpenter, memorandum to the Vice President.

31. Ken Hechler, "Extension of Remarks 'Women Can Be Astronauts,' " *Congressional Record*, March 15, 1962.

32. Bob Fenley, " 'Astro-nettes' Next? Let Cosmos Beware!" *Dallas Times Herald*, March 15, 1962.

33. "Kerr Skirts 'Astronette' Squabble," n. pub., n.d. Jerrie Cobb to James Webb attachment, May 18, 1962 (HST). Jerrie Cobb, letter to F.L.A.T.s, n.d. (IWASM).

34. Jerrie Cobb, e-mail to author, August 10, 2002.

35. Donald Ritchie (Senate historian), telephone interview with author, August 1, 2002.

36. Allan Cromley, "Women 'Race' for Space," *The Daily Oklahoman*, March 15, 1962.

37. Jane Hart, interview with author, September 26, 2001.

38. "Administrator's Memorandum on Equal Opportunity for Women," February 20, 1962 (NASA).

39. Jane Hart, interview with author, September 26, 2001; Cromley, "Women 'Race' for Space"; "Appeal to Johnson: 2 Would-Be 'Astronettes' Plead: Let Us Beat Reds," UPI, n.d. (DDE).

40. Dorothy Anderson, telephone interview with author, June 1, 2002; Donald Ritchie (Senate historian), telephone interview with author, August 1, 2002.

41. Lyndon Johnson, letter to Jacqueline Cochran, April 13, 1962 (DDE).

42. Warren G. Woodward, interview with Paul Bolton, June 3, 1968, transcript AC 69-82, 30–31 (LBJ).

43. Jerrie Cobb, letter to Lyndon Johnson, April 17, 1962 (LBJ); Jane Hart, interviews with author, September 26 and November 26, 2001; Jerrie Cobb, e-mail to author, August 10, 2002.

44. Jane Hart, interview with author, September 26, 2001.

45. Cromley, "Women 'Race' for Space."

46. Jane Hart, interview with author, September 26, 2001; Jerrie Cobb, e-mail to author, August 10, 2002.

47. Liz Carpenter, telephone interview with author, November 19, 2001.

48. Lyndon Johnson, letter to James Webb, March 15, 1962 (draft by Liz Carpenter) (LBJ).

CHAPTER TEN:
CONGRESSIONAL HEARINGS:
SOUND AND PROPER

1. Mrs. George B. Ward, Jr., letter to Lyndon Johnson, March 24, 1962 (LBJ).

2. Catherine Smith, letter to Lyndon Johnson, March 16, 1962 (LBJ).

3. "A Bachelor," letter to Lyndon Johnson, March 15, 1962 (LBJ).

4. Ashton Graybiel, letter to Jerrie Cobb, March 22, 1962 (JC).

5. Marie Smith, "Senator's Wife Wants Distaff Elbow Room: Asks Space for Women in Space," *Washington Star*, n.d. (DDE).

6. Jerrie Cobb, letter to F.L.A.T.s, n.d. (IWASM).

7. "House Group: 9 Out of 11 Men: A Probe on Discrimination Against Women in Space," *The New York Times*, June 15, 1962; "House to Probe Bias Against 'Spacewomen,' " *Daily Press Newport News*, June 15, 1962; Jacqueline Cochran, letter to Jane Hart, June 15, 1962 (DDE).

8. Helen Colson, "House Hearings on Women in Space Are Scheduled to Begin July 17," *Washington Daily News*, June 22, 1962.

9. "House to Probe Bias Against 'Spacewomen' "; "House Group: 9 Out of 11 Men: A Probe of Discrimination Against Women in Space"; Colson, "House Hearings on Women in Space Are Scheduled to Begin July 17"; "Spacewomanship: Lady Astronauts Aspirants Serious," *Detroit Free Press*, July 18, 1962.

10. Jane Hart, interview with author, November 26, 2001.

11. Jane Hart, interview with author, September 26, 2001.

12. Jacqueline Cochran, letter to Jane Hart, June 15, 1962 (DDE).

13. James Webb, letter to Jacqueline Cochran, July 12, 1962; Robert Gilruth, letter to Jacqueline Cochran, June 26, 1962; Hugh Dryden, letter to Jacqueline Cochran, June 26, 1962; Jacqueline Cochran, letter to Curtis

LeMay, June 16, 1962; Jacqueline Cochran, letter to W. Randolph Lovelace II, June 15, 1962; Jacqueline Cochran, letter to Gene Nora Stumbough, July 2, 1962 (all DDE).

14. Jerri Sloan Truhill, e-mail to author, August 11, 2002.

15. Cathryn (Walters) Liberson, telephone interview with author, August 17, 2002.

16. Jerrie Cobb, e-mail to author, June 9, 2002.

17. Scott Carpenter, telephone interview with author, June 14, 2001; John Glenn, interview with author, July 30, 2001; Cathryn (Walters) Liberson, telephone interview with author, August 17, 2002; Donald Kilgore, interview with author, August 10, 2002; Jay T. Shurley, interview with author, January 15, 2002; Jane Hart, interview with author, September 26, 2001; Jacqueline Lovelace Johnson, interview with author, January 11, 2002; Joseph Karth, telephone interview with author, August 6, 2002.

18. W. Randolph Lovelace II, letter to Jacqueline Cochran, June 29, 1962 (DDE).

19. Jerrie Cobb and Jane Rieker, *Woman into Space: The Jerrie Cobb Story* (Englewood Cliffs, N.J.: Prentice Hall, 1963), 216.

20. "Opening Statement of Jerrie Cobb Before the Special House Subcommittee Looking into the Practicability of the Training and Use of Women as Astronauts," July 17, 1962 (JC).

21. "Qualifications for Astronauts. Hearings before the Special Subcommittee on the Selection of Astronauts," Committee on Science and Astronautics, U.S. House of Representatives, July 17 and 18, 1962 (Washington, D.C.: U.S. Government Printing Office, 1962), 1.

22. Ibid.

23. Ibid., 3.

24. Jacqueline Cochran, letter to Jane Hart, June 15, 1962 (DDE); Jane Hart, letter to Jacqueline Cochran, June 22, 1962 (DDE).

25. Irene Leverton, interview with author, July 18, 2001; Bernice Steadman, interview with author, September 26, 2001.

26. Cobb and Rieker, *Woman into Space*, 214.

27. "Woman Flier Claims She Won't Give Up," *Los Angeles Times*, July 19, 1962.

28. "Qualifications for Astronauts," 20.

29. Ibid., 5.

30. Ibid., 6.

31. Ibid., 7.

32. Robert C. Toth, "Women Pilots Make Bid for a Chunk of Space," *New York Herald Tribune*, July 18, 1962.

33. "Qualifications for Astronauts," 7–8.

34. Ibid., 9.

35. Ibid.

36. Ibid.

37. Ibid., 20.

38. Ibid.; Joseph Karth, interview with author, August 6, 2002.

39. "Qualifications for Astronauts," 22.

40. Ibid.
41. James Webb, letter to Jacqueline Cochran, July 12, 1962 (DDE); Jacqueline Cochran, Copies of Statement No. 1 (DDE); Jacqueline Cochran, Copies of Statement No. 2 (DDE).
42. "Qualifications for Astronauts," 23.
43. Ibid., 24–25.
44. Ibid., 28.
45. Ibid.
46. Ibid., 71.
47. Ibid., 28.
48. Ibid., 29.
49. Jacqueline Cochran and Maryann Bucknum Brinley, *Jacqueline Cochran: An Autobiography* (New York: Bantam Books, 1987), 318.
50. "Qualifications for Astronauts," 35, 38.
51. Ibid., 38.
52. Ibid., 39.
53. John Glenn, interview with author, July 30, 2001; Scott Carpenter, telephone interview with author, June 14, 2001.
54. G. Dale Smith, letter to George Low, June 19, 1962 (NASA).
55. "Qualifications for Astronauts," 52, 53, 61, 66.
56. Ibid., 53.
57. R. B. Pirie, letter to Jacqueline Cochran, April 30, 1962 (DDE).
58. Philip Phillips, letter to Jay Shurley, September 18, 1961 (CL); Jerrie Cobb, e-mail to author, August 10, 2002.
59. Cobb and Rieker, *Woman into Space*, 217.
60. "Qualifications for Astronauts," 55; John Glenn with Nick Taylor, *John Glenn: A Memoir* (New York: Bantam Books, 2000), 247.
61. "Qualifications for Astronauts," 75.
62. Ibid., 57–58; Isabelle Shelton, "Bachelor Fulton Waxes Poetic on Behalf of Women in Space," *Washington Star*, July 18, 1962.
63. "Qualifications for Astronauts," 58.
64. Scott Carpenter, telephone interview with author, June 14, 2001.
65. "Qualifications for Astronauts," 58.
66. Ibid., 64; John Glenn, interview with author, July 30, 2001.
67. "Qualifications for Astronauts," 67.
68. Earl Lane, "No Liftoff: Women in Original Astronaut Training Program Are Still Disappointed and Angry," *Newsday*, October 27, 1998.
69. Jane Hart, interview with author, September 26, 2001.
70. "Qualifications for Astronauts," 67, 68, 69.
71. Ibid., 74.
72. Ibid., 74, 75.
73. Ibid., 77–83, 84.
74. Cathryn (Walters) Liberson, telephone interview with author, October 31, 2002; "2 Astronauts 'Scrub' Bid of Women Pilots," Chicago Tribune Press Service, July 19, 1962.

75. Gwen Gibson, "Kennedy-Glenn Team: In Hyannis Port Regatta Space Was in the Drink," *The Washington Post*, July 23, 1962.

76. Mary Walsh, "Hopes for Early Flight: Girl Scouts Hear 'Space Woman,' " *The Boston Herald*, July 26, 1962; "Girl Scouts Told Space Program Needs Women," *New York Herald Tribune*, July 26, 1962.

77. Jerrie Cobb, telegrams to John Kennedy, July 20 and August 13, 1962 (NASA); White House routing slip, July 26, 1962 (NASA).

78. "Report on the Activities of the Committee on Science and Astronautics," U.S. House of Representatives, Eighty-Seventh Congress (Washington, D.C.: U.S. Government Printing Office, 1962); "Report of the Special Subcommittee on the Selection of Astronauts: Qualifications for Astronauts," Committee on Science and Astronautics, U.S. House of Representatives, Eighty-Seventh Congress, Second Session, Serial S (Washington, D.C.: U.S. Government Printing Office, 1962).

CHAPTER ELEVEN: AFTERBURN

1. Jerrie Cobb, telegram to Bernice Steadman, July 18, 1962 (IWASM).

2. Jerrie Cobb, letter to Bernice Steadman, September 11, 1962 (IWASM); Jerrie Cobb, e-mail to author, May 7, 2002.

3. Robert B. Voas, Speech to Downtown YMCA, February 1, 1963 (NASA).

4. Betty Gillis, letter to Jean Hixson, September 19, 1962 (PV).

5. Hugh Dryden, letter to Jacqueline Cochran, June 18, 1962 (DDE).

6. Floyd Odlum, memorandum, n.d. (DDE).

7. Wernher von Braun, transcript of speech at Mississippi State College, November 19, 1962 (NASA).

8. Jay T. Shurley, telephone interview with author, May 17, 2002.

9. Jean Hixson, airline return receipt (PV).

10. W. Randolph Lovelace II, letter to Jacqueline Cochran, July 16, 1962 (DDE).

11. Johnnie R. Betson, Jr., and Robert R. Secrest, "Prospective Women Astronauts Selection Program: Rationale and Comments," *American Journal of Obstetrics and Gynecology* 88 (February 1, 1964), 421–423.

12. James Webb, letter to Jerrie Cobb, September 5, 1962 (JC).

13. Jerrie Cobb, letter to James Webb, August 7, 1962 (NASA).

14. Sue Cronk, "NASA Program Winged by Aviatrix," *The Washington Post*, November 16, 1962; Leonard Raphael, "Ballad of the Lonely Space Scientist," edited by Martin Levin (JC).

15. R. P. Young, "Memorandum for Record: Subject: Meeting Between the Administrator and Miss Jerrie Cobb," December 27, 1962 (NASA).

16. NASA, "Webb Announced Today the Appointment of Miss Jacqueline Cochran as a Special Consultant at NASA," "For Release Draft," June 11, 1962 (DDE).

17. Bernice Steadman, interview with author, September 26, 2001; Myrtle Cagle, telephone interview with author, January 2, 2002.

18. Jerri Sloan Truhill, interview with author, July 20, 2001.

19. Wally Funk, interview with author, December 4, 1998; Petra Illig, e-mail to author, November 5, 2002.

20. "Women Demand Astronaut Roles, Court Posts, Storm Name Changes," *The New Haven Register,* December 8, 1969; Jane Hart, interview with author, September 26, 2001.

21. Charles L. Sanders, "The Troubles of 'Astronaut' Edward Dwight," *Ebony,* June 2, 1965, 29–36; Joseph D. Atkinson, Jr., and Jay M. Shafritz, *The Real Stuff: A History of NASA's Astronaut Recruitment Program* (New York: Praeger, 1985), 100–104, 108, quoting "Soviets," UPI teletype no. 19, June 8, 1965.

22. Joy Miller, "Space-Bitten U.S. Gal Chagrined," *Hackensack* (N.J.) *Record,* June 17, 1963.

23. Clare Boothe Luce, "But Some People Simply Never Get the Message," *Life,* June 28, 1963, 31.

24. Ibid.

25. "Aviatrix Applies for Space Training But Is Too Late," *Newport News Daily Press,* July 12, 1963.

26. "Miss Cobb Quits Firm," *The Daily Oklahoman,* November 22, 1963.

27. Lyndon Johnson and James Webb, telephone conversation, "Recordings and Conversations and Meetings," tape K6311.03, November 26, 1963 (LBJ).

28. Ibid., 379–380; Jacqueline Lovelace Johnson, interview with author, January 11, 2002.

29. Richard G. Elliott, " 'On a Comet, Always': A Biography of W. Randolph Lovelace II," *New Mexico Quarterly* 36 (1966–1967), 376–380 (UNM).

30. Ibid., 351–352; Thomas Chiffelle, interview with author, August 20, 2000.

31. John Loosbrock, "Randy Lovelace 1907–1965," *Air Force Magazine,* January 1966, 75; Elliott, " 'On a Comet, Always,' " 350.

32. Stanley Mohler, telephone interview with author, April 17, 2002.

33. Jo Werne, "For and About Women: Jerrie Cobb," *Miami Herald,* April 18, 1962.

34. Ibid.

35. Jerrie Cobb, *Jerrie Cobb, Solo Pilot* (Sun City Center, Fla.: Jerrie Cobb Foundation, Inc., 1997), 166.

36. Francis X. Donnelly, "Pioneer Flier Shoots for Stars, Bids for Spaceflight," *Florida Today,* June 21, 1998.

37. Meg Laughlin, "The Discarded Astronaut," *Tropic* (Sunday magazine of *The Miami Herald*), June 12, 1983, 10–13, 15–17, 19–20, 22.

38. Cobb, *Jerrie Cobb: Solo Pilot,* 213.

39. Ibid., 194.

40. Stephen J. Garber and Roger D. Launius, "A Brief History of the National Aeronautics and Space Administration" (NASA History Office, Internet copy); Roger E. Bilstein, *Orders of Magnitude: A History of the NACA and NASA, 1915–1990* (Washington, D.C.: U.S. Government Printing Office, 1989) (Internet copy), chapters 6 and 7.

41. Yvonne C. Pateman, *Women Who Dared: American Female Test Pilots, Flight-*

Test Engineers, and Astronauts, 1912–1996 (Laguna Hills, Calif.: Norstahr Publishing, 1997), 76, 89, 103.

42.   Atkinson and Shafritz, *The Real Stuff,* 134–137.

43.   George Low, memorandum, August 14, 1973 (NASA).

44.   Barbara Jordan, letter to Nichelle Nichols, March 14, 1977 (NASA).

45.   Atkinson and Shafritz, *The Real Stuff,* 133.

46.   "Well-Wishers Watch Sally Ride Take New Step for Womankind," *The Washington Post,* June 19, 1983.

EPILOGUE: LEFT SEAT

1.   Jerri Sloan Truhill, interview with author, July 20, 2002; Sarah Gorelick Ratley, interview with author, January 12, 2002.

2.   Eileen Collins, Shuttle Commander Ceremony, March 6, 1998 (transcript) (Internet copy, www.flatoday.com/space/explore/stories/1998/030698e. htm).

3.   F.L.A.T.s 1986 International Women's Air and Space Museum reunion videotape (IWASM).

4.   Wally Funk personal videotape of Mercury 13 Banquet, Oklahoma City, Okla., 1994 (WF).

5.   Eileen Collins, telephone interview with author, February 2, 2000.

6.   Wally Funk, telephone interview with author, September 12, 2002.

7.   Jerri Sloan Truhill, interview with author, July 20, 2001; Sarah Gorelick Ratley, interview with author, January 12, 2002.

8.   Marcia Dunn, "38 Years Later, Jerrie Cobb Wants Her Shot at Space," *The Seattle Times,* July 13, 1998.

BIBLIOGRAPHY

"Administrator's Memorandum on Equal Opportunity for Women." February 20, 1962. (NASA)

Albuquerque Journal. Uncatalogued Lovelace Papers Collection, Box 13. (LF)

"Appeal to Johnson: 2 Would-Be 'Astronettes' Plead: Let Us Beat Reds." UPI, n.d. (DDE)

"Army Doctor's Record Parachute Jump." *Life*, August 9, 1943, 69.

Atkinson, Joseph D., Jr., and Jay M. Shafritz. *The Real Stuff: A History of NASA's Astronaut Recruitment Program*. New York: Praeger, 1985.

"Aviatrix Applies for Space Training But Is Too Late." *Newport News Daily Press*, July 12, 1963.

Baker, Kathryn Yowell. "Jerrie." *The New Classen Life* 12, no. 50 (Winter 1999), 22–23.

Battelle, Phyllis. "Blast Off, Ladies: Why Not Women in Outer Space?" *San Francisco News Call Bulletin*, July 24, 1962.

Bean, Covey. "OU Professor Misses Out on Profits of Latest Fad." *The Daily Oklahoman*, October 15, 1979.

Beschloss, Michael R. *Mayday: Eisenhower, Khrushchev, and the U-2 Affair*. New York: Harper and Row, 1986.

Betson, Johnnie R., Jr., and Robert R. Secrest. "Prospective Women Astronauts Selection Program: Rationale and Comments," *American Journal of Obstetrics and Gynecology* 88 (February 1, 1964).

Bilstein, Roger E. *Orders of Magnitude: A History of the NACA and NASA, 1915–1990*. Washington, D.C.: U.S. Government Printing Office, 1989. (Internet copy)

Braun, Wernher von. Transcript of speech at Mississippi State College, November 19, 1962. (NASA)

Burrows, William E. *Deep Black: Space Espionage and National Security*. New York: Random House, 1986.

———. *This New Ocean: The Story of the First Space Age*. New York: Modern Library, 1998.

Carl, Ann. *A WASP Among Eagles: A Woman Military Test Pilot in World War II*. Washington, D.C.: Smithsonian Institution Press, 1999.

Carpenter, Liz. Memorandum to the Vice President, "Background for Your Conference at 11 a.m. Thursday on Women in Space," March 14, 1962. (LBJ)

Carpenter, Susan. "Rocket Grrrls!" *George*, September 1997.

"CBS Special Reports," telecast, April 12, 1961. Museum of Television and Radio, New York, N.Y.

"CBS Special Reports," telecast, May 5, 1961. Museum of Television and Radio, New York, N.Y.

Cobb, Jerrie. *Jerrie Cobb: Solo Pilot*. Sun City Center, Fla.: Jerrie Cobb Foundation, Inc., 1997.

———. Report to NASA, June 15, 1961. (NASA)

———. "Woman as Astronauts." Unpublished manuscript. (JC)

———. "Woman's Participation in Space Flight." Speech to meeting of Aviation/Space Writers Association, May 1, 1961, New York City. (JC)

Cobb, Jerrie, and Jane Rieker. *Woman into Space: The Jerrie Cobb Story*. Englewood Cliffs, N.J.: Prentice Hall, 1963.

Cochran, Jacqueline. Interoffice memo, December 6, 1960. (DDE)

———. *The Stars at Noon*. Boston: Little, Brown, 1954.

———. "Women in Space: Famed Aviatrix Predicts Women Astronauts Within Six Years." *Parade*, April 30, 1961, 8.

Cochran, Jacqueline, and Maryann Bucknum Brinley. *Jacqueline Cochran: An Autobiography*. New York: Bantam Books, 1987.

Coffey, Ivy. "City Woman's Eager to Make Pioneering Flight into Space." *The Daily Oklahoman*, August 19, 1960.

———. "First Spacelady." *The Daily Oklahoman*, October 23, 1960.

———. "Jerrie Cobb Flies Jet Fighter: Pilot Adds Another 'First' to Career." *The Daily Oklahoman*, October 29, 1959.

———. "Jerrie's Story." Unpublished manuscript. (JC)

———. "Meet Jerrie Cobb—First Woman-in-Space Candidate." *The Daily Oklahoman*, n.d.

———. "Red Carpet Greets Girl Pilot." *The Daily Oklahoman*, March 21, 1959.

———. "The Story of Jerrie Cobb: First American Woman to Qualify as an Astronaut." *Guideposts*, August 1961, 1–5.

Collins, Eileen. Shuttle Commander Ceremony, March 6, 1998 (transcript). (Internet copy [www.flatoday.com/space/explore/stories/1998/030698e.htm])

Colson, Helen. "House Hearings on Women in Space Are Scheduled to Begin July 17." *Washington Daily News*, June 22, 1962.

Coogler, Edith Hills. "Possible Woman Astronaut: Tar Heel Set Sights on Moon." *Charlotte Observer*, July 14, 1963.

Cox, Donald. "Woman Astronauts." *Space World*, September 1961, 37, 58–60.

Cromley, Allan. "Women 'Race' for Space." *The Daily Oklahoman*, March 15, 1962.

Cronk, Sue. "NASA Program Winged by Aviatrix." *The Washington Post*, November 16, 1962.

Crossfield, A. Scott. NASA Biographical Data Fact Sheet. (www.nasa.gov)

"Damp Prelude to Space: A Potential Lady Orbiter Excels in Lonesome Test." *Life*, October 24, 1960, 81.

"Death Ruled a Suicide." *The New York Times*, October 20, 1960.

"Democrats Praise Kennedy Talk, Call for Bold Challenge." *The Daily Oklahoman*, January 31, 1961.

Dickson, Paul. *Sputnik: The Shock of the Century*. New York: Walker Publishing Company, 2001.

Dietrich, Marion. "First Woman into Space." *McCall's*, September 1961, 80–81, 180, 182, 184.

Donnelly, Francis X. "Pioneer Flier Shoots for Stars, Bids for Spaceflight." *Florida Today,* June 21, 1998.

"Dr. Donald D. Flickinger, 89, A Pioneer in Space Medicine." *The New York Times,* March 3, 1997.

Dunn, Marcia. "NASA Pioneer Asks for Her Shot at Space: 1st Female Astronaut Hopeful Eyes Shuttle." *The Washington Post,* July 13, 1998.

———. "38 Years Later, Jerrie Cobb Wants Her Shot at Space." *The Seattle Times,* July 13, 1998.

"DWD Jr., Other Top Douglas Officials Will Be Here to Take Part in World Space Conference." *Tulsa World,* Peaceful Uses of Space Section, May 27, 1961.

Elliott, Rand, and Daniel Carey. "Gold Dome Reflects City Heritage." *The Daily Oklahoman,* January 16, 2002.

Elliott, Richard G. " 'On a Comet, Always': A Biography of W. Randolph Lovelace II." *New Mexico Quarterly* 36 (1966–1967). (UNM)

Englund, Julie I. "First-Rate, Second-Class." *The Washington Post,* May 13, 2002.

"Experts on Space Research to Appear Here." *Tulsa World,* Peaceful Uses of Space Section, May 27, 1961.

Fenley, Bob. " 'Astro-nettes' Next? Let Cosmos Beware!" *Dallas Times Herald,* March 15, 1962.

Fields, Sidney. "The Indestructible First Lady of World Aviation." *The New York Mirror,* January 12, 1958.

"First Jet Ride." *Boston Traveler,* July 7, 1955.

"First Organized Aeromedical Research on Manned Space Flight." *Air Force Aerospace Medical Research Laboratory: 50 Years of Research on Man in Flight.* USAF, 1985 (Internet copy). (WP)

"First Woman Astronaut Trainee Tells of Program for Space." *Los Angeles Times,* February 23, 1962.

Flander, Judy. "NOW Blanks Too Hard on One Zap." *Washington Star,* August 27, 1974.

F.L.A.T.s 1986 International Women's Air and Space Museum reunion videotape. (IWASM)

"Four States Bid for '62 National Space Parlay." *Tulsa World,* May 28, 1961.

Frankman, Betty Skelton, interview by Carol L. Butler, July 19, 1999, Cocoa Beach, Florida, NASA Oral History Project. (NASA)

"From Aviatrix to Astronatrix." *Time,* August 29, 1960.

Funk, Wally. Videotape. Personal videotape of Mercury 13 Banquet, Oklahoma City, Okla., 1994. (WF)

Garber, Stephen J., and Roger D. Launius. "A Brief History of the National Aeronautics and Space Administration." (NASA History Office, Internet copy)

Gelber, Carol. "Ruth Nichols Flies Through the Years." *Philadelphia Evening Bulletin,* May 28, 1959.

Gennett, Ann. "Space for Women Too." *Contributions of Women: Aviation.* Minneapolis: Dillon Press, 1975.

Gibson, Gwen. "Kennedy-Glenn Team: In Hyannis Port Regatta Space Was in the Drink." *The Washington Post,* July 23, 1962.

"Girl Scouts Told Space Program Needs Women." New York Herald Tribune, July 26, 1962.

"Girls Ride into Space Still Long Way Off." The Dayton Daily News, September 29, 1960.

Glenn, John, with Nick Taylor. John Glenn: A Memoir. New York: Bantam Books, 2000.

Green, Richard. "The Early Years: Jolly West and the University of Oklahoma Department of Psychiatry." Journal of the Oklahoma State Medical Association 93, no. 9 (September 2000), 446–454.

Haynsworth, Leslie, and David Toomey. Amelia Earhart's Daughters: The Wild and Glorious Story of American Women Aviators from World War II to the Dawn of the Space Age. New York: Morrow, 1998.

Hechler, Ken. "Extension of Remarks 'Women Can Be Astronauts.' " Congressional Record, March 15, 1962.

Hoffstetter, Jane. "She'd Be First Woman in Space." Fort Lauderdale News, May 19, 1961.

Holden, Henry, and Lori Griffith. Ladybirds II: The Continuing Story of American Women in Aviation. Mt. Freedom, N.J.: Blackhawk Publishing Co., 1993.

"House Group: 9 Out of 11 Men: A Probe on Discrimination Against Women in Space." The New York Times, June 15, 1962.

"House to Probe Bias Against 'Spacewomen.' " Daily Press Newport News, June 15, 1962.

"It Takes Snow to Keep Publishers Away." The Daily Oklahoman, January 31, 1961.

"James Webb Chosen to Head Space Agency, Experienced as Administrator." The Daily Oklahoman, January 31, 1961.

Jerrie Cobb file. (NASA)

"Jerrie Cobb Soars to New Record in Aero Commander June 13." 99 News, July 1957. (JC)

"Jet Record Claimed." The New York Times, January 22, 1958.

Johnson, Lyndon B., and James Webb, telephone conversation. "Recordings and Conversations and Meetings," tape K6311.03, November 26, 1963. (LBJ)

Kennedy, John F. "Urgent National Needs," telecast speech before joint session of Congress, May 25, 1961. Museum of Television and Radio, New York, N.Y.

"Kerr Receiving Tribute Friday from Chamber." The Daily Oklahoman, January 21, 1961.

"Kerr Skirts 'Astronette' Squabble." n.pub., n.d. Jerrie Cobb to James Webb attachment, May 18, 1962. (HST)

Kilgore, Donald E. Oral history interview by Jake Spidle, November 8 and 15 and December 9, 1985, Oral History of Medicine Project. (UNM)

Kirby, Gordon W. "Feminine Astronauts Said Casualty of Berlin Crisis." Idaho Statesman, August 16, 1962.

Kocivar, Ben, producer. "The Lady Wants to Orbit." Look, February 2, 1960, 113.

"A Lady Proves She's Fit for Space Flight." Life, August 29, 1960, 73–75.

Lambright, W. Henry. Powering Apollo: James E. Webb of NASA. Baltimore: Johns Hopkins University Press, 1995.

Lane, Earl. "No Liftoff: Women in Original Astronaut Training Program Are Still Disappointed and Angry." *Newsday*, October 27, 1998.

Laughlin, Meg. "The Discarded Astronaut." *Tropic* (Sunday magazine of *The Miami Herald*), June 12, 1983, 10–13, 15–17, 19–20, 22.

Launius, Roger D. Introduction to *The Birth of NASA: The Diary of T. Keith Glennan*. Washington, D.C.: National Aeronautics and Space Administration, 1993.

———. *NASA: A History of the U.S. Civil Space Program*. Malabar, Fla.: Krieger Publishing Company, 1994.

"Leap from the Stratosphere." *Boeing News*, August 1943, 3–16.

"Like Man: Women Can Be 'Way Out,' Too." *The Washington Post*, May 21, 1961.

Lilly, John C., and Jay T. Shurley. "Experiments in Solitude, in Maximum Achievable Physical Isolation with Water Suspension, of Intact Healthy Persons." In *Psychophysiological Aspects of Space Flight*, edited by Bernard E. Flaherty. New York: Columbia University Press, 1961.

Lindley, Denver. "Take a Deep Breath." *Collier's*, November 23, 1940.

Link, Mae Mills. "Toward Countdown." In *Space Medicine in Project Mercury*. Washington, D.C.: National Aeronautics and Space Administration, Scientific and Technical Information Division, 1965. (Internet copy)

"List of Women Who Completed Medical Tests at Lovelace Foundation, Albuquerque, N.M." (DDE)

Loosbrock, John. "Randy Lovelace 1907–1965." *Air Force Magazine*, January 1966, 75.

"Lovelace Foundation for Medical Education and Research," *Albuquerque Journal*, n.d. (UNM)

Lovelace, Randy (W. Randolph Lovelace II). "Duckings, Probings, Checks That Proved Fliers' Fitness." *Life*, April 20, 1959.

Lovelace, W. Randolph, II. "Human Factors in Space Exploration." Unpublished manuscript. (UNM)

Lovelace, W. Randolph, II, A. H. Schwichtenberg, Ulrich C. Luft, and Robert R. Secrest. "Selection and Maintenance Program for Astronauts for the National Aeronautics and Space Administration," *Aerospace Medicine*, June 1962, 667–684.

Low, George. Memorandum, August 14, 1973. (NASA)

Luce, Clare Boothe. "But Some People Simply Never Get the Message." *Life*, June 28, 1963, 31.

Luft, Dr. Ulrich. Oral history interview by Jake Spidle. October 11 and 16, 1985, Oral History of Medicine Project. (UNM)

Martin, Edward T. "The Hero at High Altitude Flight." *Airline Pilot*, February 1983, 22–24, 38–39.

McConnell, Georgiana. "Written after returning from Albuquerque in 1961." (GM)

McCullough, Joan. "The 13 Astronauts Who Were Left Behind." *Ms.*, September 1973, 41–45.

McKee, Ruby Clayton. "Jacqueline Cochran Stresses Top Role for Women in Space." *The Dallas Morning News*, January 18, 1961.

Mercury Astronaut Selection Fact Sheet, April 9, 1959. (NASA)

Miller, Joy. "Space-Bitten U.S. Gal Chagrined." *Hackensack* (N.J.) *Record,* June 17, 1963.

"Miss Cobb Quits Firm." *The Daily Oklahoman,* November 22, 1963.

Murray, Charles, and Catherine Bly Cox. *Apollo: The Race to the Moon.* New York: Simon and Schuster, 1989.

NASA. Memorandum, "Proposed Answers to Female Astronaut Volunteers," May 31, 1961. (NASA)

NASA. "Webb Announced Today the Appointment of Miss Jacqueline Cochran as a Special Consultant at NASA." "For Release Draft," June 11, 1962. (DDE)

NASA. Press release, April 9, 1959. (NASA)

Newcomb, Harold. "Cochran's Convent." *Airman,* May 20, 1977, n.p. (NASA)

Nichols, Ruth. "Why Not Lady Astronauts?" *Washington Daily News,* November 24, 1959.

Nichols, Ruth. *Wings for Life.* Philadelphia: Lippincott, 1957.

Oklahoma College for Women Catalogue, 1949.

"1,596 Watch Queens Edge Phoenix, 3-1." *The Daily Oklahoman,* May 10, 1947.

"Opening Statement of Jerrie Cobb Before the Special House Subcommittee Looking into the Practicability of the Training and Use of Women as Astronauts," July 17, 1962. (JC)

Pateman, Yvonne C. *Women Who Dared: American Female Test Pilots, Flight-Test Engineers, and Astronauts, 1912–1996.* Laguna Hills, Calif.: Norstahr Publishing, 1997.

"Police Suspect Suicide: Ruth R. Nichols, Famed Flier Dies in Apartment." *The Boston Globe,* September 26, 1960.

"Press Conference Mercury Astronaut Team" transcript. April 9, 1959. (NASA)

"Qualifications for Astronauts. Hearings before the Special Subcommittee on the Selection of Astronauts." House Committee on Science and Astronautics, U.S. House of Representatives, July 17 and 18, 1962. 87th Congress, 2nd Session. Washington, D.C.: U.S. Government Printing Office, 1962.

"Reminiscences of Ruth Nichols." Interview by Kenneth Leish. June 1960. In the Oral History Collection, Part IV (1-219), Columbia University, New York, N.Y.

Renner, Tom. "Aviatrix Flies 1,000 MPH over LI, Sets New Mark." *Newsday,* January 22, 1958.

"Report on the Activities of the Committee on Science and Astronautics." U.S. House of Representatives, 87th Congress. Washington, D.C.: U.S. Government Printing Office, 1962.

"Report of the Special Subcommittee on the Selection of Astronauts: Qualifications for Astronauts." House Committee on Science and Astronautics. U.S. House of Representatives, 87th Congress, 2nd Session, Serial S. Washington, D.C.: U.S. Government Printing Office, 1962.

Rieker, Jane. "Up and Up Goes Jerrie Cobb." *Sports Illustrated,* August 29, 1960, 28–29.

Roe, Dorothy. "Space Is Goal of Gal Flyer." *The Baltimore Evening Sun,* February 14, 1958.

Ruff, George E., and Edwin Z. Levy. "Psychiatric Evaluation of Candidates for Space Flight." *The Journal of Psychiatry* 116, no. 5 (November 1959), 385–391.

"Ruth Nichols Death Here Listed by Police as Possible Suicide." *The New York Times,* September 26, 1960.

"Ruth Nichols Found Dead, May Have Been a Suicide." *Boston Herald Tribune,* September 27, 1960.

"Ruth Nichols Rivals Amelia Earhart in Aviation Accomplishments." *Foundation for the Carolinas Newsletter,* Winter 1999.

Sanders, Charles L. "The Troubles of 'Astronaut' Edward Dwight." *Ebony,* June 2, 1965, 29–32, 34–36.

Schwichtenberg, A. H. Oral history interview by Jake Spidle, February 20, 1985, Oral History of Medicine Project. (UNM)

Secrest, Dr. Robert. Oral history interview by Jake Spidle, July 8, 1996, Oral History of Medicine Project. (UNM)

"Seven Brave Women Behind the Astronauts." *Life,* September 21, 1959, 142–163.

"She Orbits over the Sex Barrier: Blue-Eyed Blond with a New Hairdo Stars in a Russian Space Spectacular." *Life,* June 28, 1963.

Shelton, Isabelle. "Bachelor Fulton Waxes Poetic on Behalf of Women in Space." *Washington Star,* July 18, 1962.

Shurley, Jay T. "Cost Ceiling Estimate." (JS)

———. "The Hydro-Hypodynamic Environment." In *Proceedings of the Third World Congress of Psychiatry,* vol. 3. Montreal: McGill University Press, 1961.

———. "Mental Images in Profound Experimental Sensory Isolation." In *Hallucinations,* edited by Louis Jolyon West. New York: Grune & Stratton, 1962.

———. Edited copy of NASA Press Release no. 59-113, "Mercury Astronaut Selection Fact Sheet." (JS)

———. "Profound Experimental Sensory Isolation." *The American Journal of Psychiatry* 117, no. 6 (December 1960), 539–545.

———. Project Venus files. (JS)

Shurley, Jay T., and Cathryn Walters. "Women Astronaut Assessment in Hydrohypodynamic Environment," August 8, 1961. (JS, JC)

Sis, Frank. "Say Astronauts to Get Company." *The Cleveland Press,* July 18, 1961.

Smith, Marie. "Senator's Wife Wants Distaff Elbow Room: Asks Space for Women in Space." *Washington Star,* n.d. (DDE)

"Softball Back, 'N with Curves." *The Daily Oklahoman,* May 10, 1947.

Soldan, Ray. "Retirement Can't Stop Jim Conger." *Oklahoma City Times,* June 29, 1965.

"Space Future for Women: Test Foundation to Be Expanded." *Tulsa World,* May 27, 1961.

"The Spaceman's Wife: 'Alan Was in His Right Place.' " *Life,* May 12, 1961, 28–29.

"Spacewomanship: Lady Astronauts Aspirants Serious." *Detroit Free Press,* July 18, 1962.

Spidle, Jake W., Jr. *The Lovelace Medical Center: Pioneer in American Health Care.* Albuquerque: University of New Mexico Press, 1987.

————. *The Lovelace Medical Center: Toward the 21st Century.* Albuquerque: University of New Mexico Press, 1992.

"*The Star,*" April 14, 1959. (NASA)

"Star Trek Star Recruits Women for NASA Program." *Palo Alto Times,* June 22, 1977.

Steadman, Bernice Trimble, with Jody M. Clark. *Tethered Mercury: A Pilot's Memoir, The Right Stuff . . . But the Wrong Sex.* Traverse City, Mich.: Aviation Press, 2001.

Swenson, Lloyd S., James M. Grimwood, and Charles C. Alexander. *The New Ocean: A History of Project Mercury.* NASA SP-4201. Washington, D.C.: U.S. Government Printing Office, 1966.

"The Tank." Videotape. WKY-TV, Oklahoma City, 1960. (JS)

Thomas, Shirley. "Don D. Flickinger: With Zest and Dedication, This Energetic Doctor Has Long Concentrated on the Problems of Man's Survival in the Hostile Environment of Space." *Men of Space: Profiles of the Leaders in Space Research, Development, and Exploration,* vol. 3. Philadelphia: Chilton Company, 1961.

————. "William Randolph Lovelace II." *Men of Space: Profiles of the Leaders in Space Research, Development, and Exploration,* vol. 2. Philadelphia: Chilton Company, 1961.

Toth, Robert C. "Women Pilots Make Bid for a Chunk of Space." *New York Herald Tribune,* July 18, 1962.

"Toward the Endless Frontier." Hearings of the Committee on Science and Technology 1959–79. U.S. House of Representatives. Washington, D.C.: U.S. Government Printing Office, 1980.

"12 Women to Take Astronaut Test." *The New York Times,* January 26, 1961.

"2 Astronauts 'Scrub' Bid of Women Pilots." Chicago Tribune Press Service, July 19, 1962.

"Under Sheltering Wings: Secluded Siesta Hills Area Enjoys Sounds of Air Traffic," *Albuquerque Journal,* December 28, 1961.

U.S. Congress. House Committee on Science and Astronautics. "Qualifications for Astronauts. Hearings before the Special Subcommittee on the Selection of Astronauts." 87th Congress, 2nd Session, July 17 and 18, 1962. Washington, D.C.: U.S. Government Printing Office, 1962.

U.S. Congress. House Committee on Science and Astronautics, "Qualifications for Astronauts. Report of the Special Subcommittee on the Selection of Astronauts." 87th Congress, 2nd Session, Serial S, July 17 and 18, 1962. Washington, D.C.: U.S. Government Printing Office, 1962.

"The U.S. Team Is Still Warming Up the Bench." *Life,* June 28, 1963.

Voas, Robert B. Speech to Downtown YMCA, February 1, 1963. (NASA)

Walsh, Mary. "Hopes for Early Flight: Girl Scouts Hear 'Space Woman.' " *The Boston Herald,* July 26, 1962.

Walters, Cathryn, Jay T. Shurley, and Oscar A. Parsons. "Differences in Male and Female Responses to Underwater Sensory Deprivation: An Exploratory Study." *Journal of Nervous and Mental Disease* 135, no. 4, 302–310. (JS)

Waterhouse, Helen. "Ohio Teacher Crashes Sound Barrier in Jet." *The Christian Science Monitor,* March 21, 1957.

Webb, James. Oral history interview by T. H. Baker, April 29, 1969 (Internet copy). (LBJ)

Weitekamp, Margaret. "The Right Stuff, The Wrong Sex: The Science, Culture, and Politics of the Lovelace Woman in Space Program, 1959–1963." Ph.D. dissertation, Cornell University, May 2001.

"Well-Wishers Watch Sally Ride Take New Step for Womankind." *The Washington Post,* June 19, 1983.

Werne, Jo. "For and About Women: Jerrie Cobb." *Miami Herald,* April 18, 1962.

Wheat, Chuck. "Governor Late, Kerr Tired—But Space Meet Launched." *Tulsa World,* May 26, 1961.

Wolfe, Tom. *The Right Stuff.* New York: Farrar, Straus, and Giroux, 1979.

"Woman Flier Claims She Won't Give Up." *Los Angeles Times,* July 19, 1962.

"Woman-in-Space Program Urged." *The Washington Post,* n.d. (DDE)

"A Woman Passes Tests Given to 7 Astronauts." *The New York Times,* August 19, 1960.

"Women Best Suited for Space, Pioneer Aviatrix Says." *The Washington Post,* April 16, 1960.

"Women Demand Astronaut Roles, Court Posts, Storm Name Changes." *The New Haven Register,* December 8, 1969.

"Women May Take Space Flights and 12 Are Ready." *The Kansas City Star,* n.d. (SGR)

"The Wrong Stuff?" *Dateline NBC,* February 10, 1995.

Yeager, Chuck, and Leo Janos. *Yeager.* New York: Bantam, 1985.

Young, R. P. "Memorandum for Record: Subject: Meeting between the Administrator and Miss Jerrie Cobb," December 27, 1962. (NASA)

Young, Warren. "What It's Like to Fly in Space." *Life,* April 13, 1959, 133–134, 137–139, 143–144, 147–148.

Index